T0260035

ANNALS OF MATHEMATICS STUDIES

NUMBER 16

ANNALS OF MATHEMATICS STUDIES

Edited by Emil Artin and Marston Morse

1. Algebraic Theory of Numbers, *by* HERMANN WEYL
3. Consistency of the Continuum Hypothesis, *by* KURT GÖDEL
6. The Calculi of Lambda-Conversion, *by* ALONZO CHURCH
7. Finite Dimensional Vector Spaces, *by* PAUL R. HALMOS
10. Topics in Topology, *by* SOLOMON LEFSCHETZ
11. Introduction to Nonlinear Mechanics, *by* N. KRYLOFF *and* N. BOGOLIUBOFF
14. Lectures on Differential Equations, *by* SOLOMON LEFSCHETZ
15. Topological Methods in the Theory of Functions of a Complex Variable, *by* MARSTON MORSE
16. Transcendental Numbers, *by* CARL LUDWIG SIEGEL
17. Problème Général de la Stabilité du Mouvement, *by* M. A. LIAPOUNOFF
19. Fourier Transforms, *by* S. BOCHNER *and* K. CHANDRASEKHARAN
20. Contributions to the Theory of Nonlinear Oscillations, Vol. I, *edited by* S. LEFSCHETZ
21. Functional Operators, Vol. I, *by* JOHN VON NEUMANN
22. Functional Operators, Vol. II, *by* JOHN VON NEUMANN
23. Existence Theorems in Partial Differential Equations, *by* DOROTHY L BERNSTEIN
24. Contributions to the Theory of Games, Vol. I, *edited by* H. W. KUHN *and* A. W. TUCKER
25. Contributions to Fourier Analysis, *by* A. ZYGMUND, W. TRANSUE, M. MORSE, A. P. CALDERON, *and* S. BOCHNER
26. A Theory of Cross-Spaces, *by* ROBERT SCHATTEN
27. Isoperimetric Inequalities in Mathematical Physics, *by* G. POLYA *and* G. SZEGO
28. Contributions to the Theory of Games, Vol. II, *edited by* H. KUHN *and* A. W. TUCKER
29. Contributions to the Theory of Nonlinear Oscillations, Vol. II, *edited by* S. LEFSCHETZ
30. Contributions to the Theory of Riemann Surfaces, *edited by* L. AHLFORS *et al.*

TRANSCENDENTAL NUMBERS

By Carl Ludwig Siegel

PRINCETON
PRINCETON UNIVERSITY PRESS
1949

LONDON: GEOFFREY CUMBERLEGE, OXFORD UNIVERSITY PRESS

PRINTED IN THE UNITED STATES OF AMERICA

PREFACE

This booklet reproduces with slight changes a course of lectures delivered in Princeton during the Spring term 1946. It would be misleading to call it a theory of transcendental numbers, our knowledge concerning transcendental numbers being narrowly restricted. The text deals with a few special transcendency problems of some interest, but it is more than a mere collection of scattered examples, since it involves a method which might be useful in the search of more general results.

Carl Ludwig Siegel.

April, 1949
Princeton, New Jersey.

CONTENTS

CHAPTER I

THE EXPONENTIAL FUNCTION

The most widely known result on transcendental numbers is the transcendency of π proved by Lindemann in 1882. His method is based on Hermite's previous work who discovered the transcendency of e in 1873. Both results are contained in the general Lindemann-Weierstrass theorem which will be proved in §12. We shall start with some simpler problems, namely the irrationality of e and π and related questions.

§1. The irrationality of e

The usual proof of the irrationality of e runs as follows. From the series

$$e = \sum_{k=0}^{\infty} \frac{1}{k!}$$

we get the decomposition

$$e = s_n + r_n, \quad s_n = \sum_{k=0}^{n} \frac{1}{k!}, \quad r_n = \sum_{k=n+1}^{\infty} \frac{1}{k!}$$

$$(n=1,2,\ldots).$$

Since

$$r_n = \frac{1}{(n+1)!}\left(1 + \frac{1}{n+2} + \frac{1}{(n+2)(n+3)} + \cdots\right) < \frac{e-1}{(n+1)!},$$

we find that

$$e = s_1 + r_1 < 2 + \frac{e-1}{2}, \quad e < 3;$$

therefore

$$0 < r_n < \frac{2}{(n+1)!}.$$

Put

$$n!\, s_n = a_n, \quad n!\, r_n = b_n,$$

then the number a_n is integral and

$$0 < b_n < \frac{2}{n+1} \leq 1$$

for $n = 1, 2, \ldots$. This proves that $n!\, e = a_n + b_n$ and, a fortiori, $n\, e$ is never an integer. In other words, e is irrational.

The proof is still simpler, if we use the series for e^{-1} instead of e. Then

$$e^{-1} = \sigma_n + \rho_n, \; \sigma_n = \sum_{k=0}^{n} \frac{(-1)^k}{k!}, \quad \rho_n = \sum_{k=n+1}^{\infty} \frac{(-1)^k}{k!}$$

$$(n=1, 2, \ldots)$$

and

$$0 < (-1)^{n+1} \rho_n = \frac{1}{(n+1)!} - \frac{1}{(n+2)!} + \ldots < \frac{1}{(n+1)!}.$$

Defining

$$n!\, \sigma_n = \alpha_n, \quad n!\, \rho_n = \beta_n$$

we see that α_n is integral and

$$0 < (-1)^{n+1}\beta_n < \frac{1}{n+1} < 1.$$

Therefore $n!e^{-1} = \alpha_n + \beta_n$ and, a fortiori, $n!e^{-1}$ is never an integer.

We can prove a little more, namely that e is not the root of a quadratic equation $ax^2 + bx + c = 0$ with integral a, b, c, not all 0. Consider the expression

$$E_n = n!\,(ae + ce^{-1})$$

with integral a and c, not both 0. Then

$$E_n = S_n + R_n, \quad S_n = aa_n + c\alpha_n, \quad R_n = ab_n + c\beta_n,$$

where S_n is integral and the absolute value

$$|R_n| \leq |ab_n| + |c\beta_n| < \frac{2|a|+|c|}{n+1},$$

so that

$$|R_n| < 1$$

for all $n \geq 2|a|+|c|$. On the other hand we have the recursion formula

$$nR_{n-1} - R_n = a(nb_{n-1} - b_n) + c(n\beta_{n-1} - \beta_n)$$

$$= a + (-1)^n c.$$

It follows that at least one of the three numbers R_{n-1}, R_n, R_{n+1} is different from 0, since otherwise $a+c=0$, $a-c=0$ and $a=0$, $c=0$. This shows the existence of a positive integer ν such that E_ν is not integral, and therefore the number

$$b + \frac{E\nu}{\nu!} = ae + b + ce^{-1}$$

is different from 0, for all integral b. This means that

$$ae^2 + be + c \neq 0$$

for arbitrary integers a, b, c, not all 0. In other words, e is not a quadratic irrationality.

§2. The operator f(D)

We denote by D the differentiation with respect to the variable x. If

$$f(t) = a_0 + a_1 t + a_2 t^2 + \dots$$

is a power series with real or complex coefficients $a_0, a_1, \dots$ and $\varphi = \varphi(x)$ a function of x, we define

$$f(D)\varphi = \sum_{n=0}^{\infty} a_n D^n \varphi = \sum_{n=0}^{\infty} a_n \frac{d^n\varphi}{dx^n}. \tag{1}$$

In order to avoid questions of convergence and differentiability, we shall apply the operator f(D) only in two cases: either φ is a polynomial, or f is a polynomial and φ possesses derivatives of all orders. In both cases, the series (1) is finite.

It is clear that, for two power series $f_1(t)$ and $f_2(t)$,

$$\Big(f_1(D) + f_2(D)\Big)\varphi = f_1(D)\varphi + f_2(D)\varphi$$

$$\Big(f_1(D)f_2(D)\Big)\varphi = f_1(D)\Big(f_2(D)\varphi\Big). \tag{2}$$

Moreover, if $a_0 \neq 0$, then

$$f^{-1}(t) = b_0 + b_1 t + \dots \qquad (b_0 = a_0^{-1})$$

exists and, by (2),

$$f^{-1}(D)\Big(f(D)\varphi\Big)=\varphi$$

for polynomials φ.

If the power series f and the polynomial φ have integral coefficients, then also $f(D)\varphi$ does, and the same holds for $f^{-1}(D)\varphi$, in case $a_0 = \pm 1$.

We define

$$J\varphi = \int_0^x \varphi(t)dt,$$

so that

$$DJ\varphi = \varphi(x), \qquad JD\varphi = \varphi(x) - \varphi(0)$$

and

(3) $$J^{n+1}\varphi = \int_0^x \frac{(x-t)^n}{n!}\varphi(t)dt$$

$$(n = 0,1,\dots).$$

In particular we are interested in the case $\varphi = e^{\lambda x}P$, where λ is a constant and $P = P(x)$ a polynomial. Since

$$D(e^{\lambda x}P) = e^{\lambda x}(\lambda + D)P,$$

we have

(4) $$D^n(e^{\lambda x}P) = e^{\lambda x}(\lambda + D)^n P \qquad (n=1,2,\dots)$$

and

(5) $$(\lambda + D)^n P = Q(x)$$

again is a polynomial. Vice versa, if $\lambda \neq 0$ and a polynomial Q are arbitrarily given, then (5) implies

$$P = (\lambda + D)^{-n} Q$$

and this is the unique polynomial solution of the differential equation

$$D^n (e^{\lambda x} P) = e^{\lambda x} Q.$$

In case $\lambda = \pm 1$, the polynomial P has integral coefficients if Q has.

§3. Approximation to e^x by rational functions

We are going to determine two polynomials $A = A(x)$, $B = B(x)$ of degree n, such that the sum $e^x + \frac{A}{B}$ vanishes at $x = 0$ of order $2n + 1$. This condition implies

$$Be^x + A = R = cx^{2n+1} + \ldots , \tag{6}$$

where $R = R(x)$ is a power series starting with a term of order $2n + 1$. Writing A and B with indeterminate coefficients and equating the terms of order 0, 1, ... , 2n in (6), we obtain $2n+1$ homogeneous linear equations for the $2n+2$ unknown coefficients in A and B. This proves that (6) has a non-trivial solution A, B. It will turn out that $c \neq 0$.

In order to obtain an explicit formula for A and B, we differentiate (6) $n+1$ times; then

$$D^{n+1}(Be^x) = D^{n+1} R$$

$$e^x (1+D)^{n+1} B = D^{n+1} R = c_0 x^n + \ldots , \tag{7}$$

where

$$c_0 = (2n+1)\ \dots\ (n+1)c.$$

Therefore

$$(8)\qquad (1+D)^{n+1}B = e^{-x}(c_0x^n+\dots) = c_0x^n+\dots = c_0x^n,$$

because $(1+D)^{n+1}B$ is a polynomial of degree n at most, and

$$B = c_0(1+D)^{-n-1}x^n.$$

To find A, we get in the same manner

$$D^{n+1}(Ae^{-x}) = D^{n+1}(Re^{-x})$$

$$e^{-x}(-1+D)^{n+1}A = c_0x^n+\ \dots$$

$$(-1+D)^{n+1}A = e^x(c_0x^n+\dots) = c_0x^n + \dots = c_0x^n$$

$$A = c_0(-1+D)^{-n-1}x^n.$$

This proves that A and B are unique, up to the arbitrary constant factor c_0. Choose $c_0=1$, then

$$(9)\qquad A(x) = (-1+D)^{-n-1}x^n, \qquad B(x) = (1+D)^{-n-1}x^n$$

have integral coefficients. Moreover, by (7) and (8),

$$D^{n+1}R = x^ne^x.$$

Since R and its first n derivatives are 0 at $x \neq 0$, we obtain, by (3),

$$R = J^{n+1}D^{n+1}R = \frac{1}{n!}\int_0^x (x-t)^n t^n e^t dt$$

(10) $$R(x) = \frac{x^{2n+1}}{n!}\int_0^1 t^n(1-t)^n e^{tx}dt$$

for n=0,1,.... Replacing t by 1-t we find

$$R(x) = \frac{x^{2n+1}}{n!}\int_0^1 t^n(1-t)^n e^{(1-t)x}dt,$$

so that

(11) $$R(x) = \frac{x^{2n+1}}{n!} e^{\frac{x}{2}} \int_0^1 t^n(1-t)^n \cosh\{(t-\tfrac{1}{2})x\}dt$$

$$(n = 0, 1, \ldots).$$

§4. <u>The irrationality of</u> e^a <u>for rational</u> $a \neq 0$

By (10),

(12) $$|R(x)| \leq \frac{|x|^{2n+1}}{n!} e^{|x|},$$

for all complex x, and

(13) $$R(x) > 0,$$

for all positive x.

Now let $x = m$ be a positive integer; then the numbers $A(m)$ and $B(m)$ are integers. Suppose that e^m were rational and denote by $q > 0$ the denominator of e^m. By (6), the number

$$qR(m) = r$$

is integral. Because of (12) and (13),

$$0 < r \leq q\frac{m^{2n+1}}{n!} e^m = q m e^m \frac{m^{2n}}{n!} < 1$$

for all sufficiently large n. This is a contradiction.

Therefore all powers e^m $(m=1,2,\dots)$ are irrational. If a is any rational number $\neq 0$, we write $a=\frac{m}{r}$ with integral $m>0$, $r\neq 0$. Since $e^m=(e^a)^r$, it follows that e^a is irrational, for all rational $a\neq 0$.

§5. The irrationality of π

We know that the polynomials $A(x)$ and $B(x)$ of degree n are uniquely determined by the formula

$$(14) \qquad B(x)e^x + A(x) = R(x) = cx^{2n+1} + \dots,$$

where $c\neq 0$ is given. Replace x by $-x$ and multiply by e^x; then

$$A(-x)e^x + B(-x) = e^x R(-x) = -cx^{2n+1} + \dots,$$

whence

$$(15) \qquad A(-x) = -B(x).$$

This can also be proved by using the expressions (9) for A and B.

Choose $x=\pi i$ and apply (11), (14) and (15); then

$$e^{\frac{x}{2}} = i, \ \cosh\{(t-\tfrac{1}{2})x\} = \cos\{(t-\tfrac{1}{2})\pi\} = \sin\pi t$$

$$(16) \qquad A(\pi i) + A(-\pi i) = R(\pi i)$$

$$(17) \qquad R(\pi i) = (-1)^{n+1}\frac{\pi^{2n+1}}{n!}\int_0^1 t^n(1-t)^n \sin\pi t \, dt.$$

The integrand is positive in the interval $0 < t < 1$, so that

$$R(\pi i) \neq 0. \tag{18}$$

The function $A(x) + A(-x)$ is a polynomial of degree $\nu = [\frac{n}{2}]$ in the variable x^2 with integral coefficients. If π^2 is a rational number and $q > 0$ its denominator, then the number

$$q^\nu R(\pi i) = j$$

is integral, by (16). However, it follows from (17) and (18) that

$$0 < |j| \leq \frac{q^{\frac{n}{2}} \pi^{2n+1}}{n!} < 1$$

for all sufficiently large values of n.

This contradiction proves that π^2 and, a fortiori, π itself are irrational.

§6. The irrationality of tg a for rational $a \neq 0$

Define

$$A(x) - A(-x) = x\, P(x^2), \qquad A(x) + A(-x) = Q(x^2),$$

so that $P=P(x^2)$, $Q=Q(x^2)$ are polynomials in x^2 of degrees $[\frac{n-1}{2}]$, $[\frac{n}{2}] = \nu$, with integral coefficients, and

$$2A(x) = Q+xP, \qquad 2A(-x) = Q-xP.$$

Multiply (14) by $e^{-\frac{x}{2}}$ and use (11), (15); then

$$A(x)\, e^{-\frac{x}{2}} - A(-x) e^{\frac{x}{2}} = R(x)\, e^{-\frac{x}{2}}$$

(19)
$$xP\cosh\frac{x}{2} - Q\sinh\frac{x}{2} = R(x)e^{-\frac{x}{2}}$$
$$= \frac{x^{2n+1}}{n!}\int_0^1 t^n(1-t)^n\cosh\{(t-\tfrac{1}{2})x\}\,dt = S(x),$$

say.

Now let $a^2=\pm b$ be a rational number $\neq 0$ and assume that also $\frac{\text{tgh}\,a}{a}$ is rational. Put $x=2a$ and denote by $q > 0$ the denominator of $x^2=4a^2=\pm 4b$. Then

$$x\cosh\frac{x}{2} = \gamma r, \qquad \sinh\frac{x}{2} = \gamma s,$$

where r, s are integers and $\gamma \neq 0$. The numbers $q^\nu P(x^2)$, $q^\nu Q(x^2)$ are integral, and (19) shows that also

$$\gamma^{-1} q^\nu S(x) = j$$

is an integer, whereas

$$|j| \leq |\tfrac{x}{\gamma}|e^{\frac{|x|}{2}}\ \frac{q^{\frac{n}{2}}|x|^{2n}}{n!} \longrightarrow 0 \quad (n\longrightarrow\infty).$$

We obtain the desired contradiction if we can prove that $R(x)\neq 0$. This inequality is obvious in case $x^2 \geq -\pi^2$, since then the integrand in (19) is positive for $0 < t < 1$. In order to complete the proof in the remaining case, we write more explicitly $A=A_n$, $B=B_n$, $R=R_n$, $c=c_n$; then

$$B_n e^x + A_n = R_n = c_n x^{2n+1} + \ldots, \quad B_{n-1}e^x + A_{n-1} = R_{n-1}$$
$$= c_{n-1}x^{2n-1} + \ldots$$
$$(n = 1, 2, \ldots)$$

$$(20)\qquad A_{n-1}B_n - A_nB_{n-1} = R_{n-1}B_n - R_nB_{n-1}$$

$$= c_{n-1}B_n(0)x^{2n-1} + \ldots = c_{n-1}B_n(0)x^{2n-1},$$

because $A_{n-1}B_n - A_nB_{n-1}$ is a polynomial in x of degree $2n-1$. If $B_n(0) = 0$, then also $A_n(0) = B_n(0) - R_n(0) = 0$, and the formula

$$(B_{n-1}+x^{-1}B_n)e^x + (A_{n-1}+x^{-1}A_n) = c_{n-1}x^{2n-1} + \ldots$$

would contradict the unicity of A_{n-1}, B_{n-1}. Therefore $B_n(0)\neq 0$. This also follows from $B(x)=(1+D)^{-n-1}x^n$, namely $B_n(0)=\binom{-n-1}{n}n!\neq 0$.

Now we infer from (20) that $R_{n-1}B_n - R_nB_{n-1}\neq 0$ for all $x\neq 0$, so that at least one of the two numbers $R_{n-1}(x)$, $R_n(x)$ is different from 0. This suffices for the completion of the proof.

Separating the real and imaginary case, we see that $\operatorname{tgh}\sqrt{b}/\sqrt{b}$ and $\operatorname{tg}\sqrt{b}/\sqrt{b}$ are irrational, for all rational positive numbers b. In particular, tg a is irrational for all rational $a\neq 0$. This contains the irrationality of π, since $\operatorname{tg}\frac{\pi}{4} = 1$ is rational.

§7. The function $P_1e^{\rho_1x} + \ldots + P_me^{\rho_mx}$

We are going to solve the following problem: Let m different complex constants $\rho_1,\ldots,\rho_m$ and m non-negative integers $n_1,\ldots,n_m$ be given and put

$$(21)\qquad \sum_{k=1}^{m}(n_k+1) = N + 1 ;$$

to determine m polynomials $P_1(x),\ldots, P_m(x)$ of degrees $n_1,\ldots,n_m$ such that the function

(22) $$R = P_1 e^{\rho_1 x} + \dots + P_m e^{\rho_m x}$$

vanishes at $x=0$ of order N. For the special case $m=2$, $n_1=n_2$, $\rho_1=1$, $\rho_2=0$, the solution of this problem was given in §3.

Writing $P_1, \dots, P_m$ with indeterminate coefficients and considering the terms of degree $0, 1, \dots, N-1$ in the power series expansion of R, we obtain N homogeneous linear equations for the $(n_1+1)+\dots+(n_m+1)$ unknown coefficients. These equations have a non-trivial solution, because of (21). This proves the existence of m polynomials $P_k(x)$ $(k=1,\dots,m)$, <u>not all identically</u> 0, of degrees $\leq n_k$, such that the power series of R takes the form

(23) $$R = c\frac{x^N}{N!} + \dots$$

with some constant c.

In case $m=1$ the solution clearly is

$$P_1 = c\frac{x^{n_1}}{n_1!},$$

so that $c \neq 0$. It will turn out that, for every m, the constant c in (23) is different from 0 and that, for any given $c \neq 0$, the solution is unique.

Write more explicitly $N=N_m$ and suppose $m > 1$. By (4) and (22),

(24) $$D^{n_m+1}(Re^{-\rho_m x}) = \sum_{k=1}^{m} D^{n_m+1}(e^{(\rho_k-\rho_m)x}P_k)$$

$$= \sum_{k=1}^{m-1} e^{(\rho_k-\rho_m)x}(\rho_k-\rho_m+D)^{n_m+1}P_k;$$

by (23),

$$(25)\qquad D^{n_m+1}(Re^{-\rho_m x}) = cD^{n_m+1}\left(\frac{x^{N_m}}{N_m!}e^{-\rho_m x}\right)$$

$$= c\frac{x^{N_{m-1}}}{N_{m-1}!} + \dots .$$

Since the m-1 polynomials

$$Q_k = (\rho_k - \rho_m + D)^{n_m+1} P_k \qquad (k=1,\dots,m-1)$$

have exactly the same degrees as P_k, they do not all vanish identically. It follows from (24) and (25) that the function

$$S = D^{n_m+1}(Re^{-\rho_m x}) = Q_1 e^{(\rho_1-\rho_m)x} + \dots + Q_{m-1}e^{(\rho_{m-1}-\rho_m)x}$$

solves the problem for m-1 and $\rho_1-\rho_m,\dots,\rho_{m-1}-\rho_m$ instead of m and $\rho_1,\dots,\rho_m$, with the same constant c. We now obtain

$$P_k = (\rho_k - \rho_m + D)^{-n_m-1} Q_k \qquad (k=1,\dots,m-1)$$

and, because of (3),

$$(26)\quad R = e^{\rho_m x} J^{n_m+1} S = e^{\rho_m x}\int_0^x \frac{(x-t)^{n_m}}{n_m!} S(t)\,dt.$$

Applying induction with respect to m, we see that we can prescribe c=1 and that then

$$P_1 = \prod_{l=2}^{m} (\rho_1 - \rho_l + D)^{-n_l-1} \frac{x^{n_1}}{n_1!},$$

more generally

(27) $$P_k = \prod_{\substack{l=1\\ l\neq k}}^{m} (\rho_k - \rho_l + D)^{-n_l-1} \frac{x^{n_k}}{n_k!} \qquad (k=1,\dots,m).$$

In order to determine R explicitly we consider first the case m=2. By (26),

$$R = e^{\rho_2 x}\int_0^x \frac{(x-t)^{n_2}}{n_2!} e^{(\rho_1-\rho_2)t} \frac{t^{n_1}}{n_1!}\,dt$$

$$= \int_{\substack{t_1+t_2=x\\ t_1>0,t_2>0}} \frac{t_1^{n_1} t_2^{n_2}}{n_1!\,n_2!} e^{\rho_1 t_1+\rho_2 t_2}\,dt_1 \qquad (x>0).$$

Suppose that the formula

(28) $$R(x) = \int\dots\int_{\substack{t_1+\dots+t_m=x\\ t_1>0,\dots,t_m>0}} \prod_{k=1}^{m} \left(\frac{t_k^{n_k}}{n_k!} e^{\rho_k t_k}\right) dt_1\dots dt_{m-1} \qquad (x>0)$$

is true for m-1≥1, instead of m. Then

$$S(t) = \int\dots\int_{\substack{t_1+\dots+t_{m-1}=t\\ t_1>0,\dots,t_{m-1}>0}} \prod_{k=1}^{m-1} \left(\frac{t_k^{n_k}}{n_k!} e^{(\rho_k-\rho_m)t_k}\right) dt_1\dots dt_{m-2} \qquad (t>0).$$

Substituting in (26) and defining $t_m=x-t$, we obtain (28) for m.

We shall call $R=P_1e^{\rho_1 x}+\dots+P_m e^{\rho_m x}$ an <u>approximation form</u>.

§8. Estimation of R(1).

From (28) we can draw two important conclusions: For arbitrary complex $\rho_1, \ldots, \rho_m$ we have the upper estimate

$$(29) \qquad |R(1)| \leq \frac{e^{|\rho_1|+\ldots+|\rho_m|}}{n_1! \ldots n_m!},$$

and for real $\rho_1, \ldots, \rho_m$ we have the lower estimate

$$(30) \qquad R(1) > 0.$$

§9. Estimation of $P_k(1)$ and its denominator

We apply the expansion

$$(31) \quad (\omega + D)^{-n-1} = \omega^{-n-1} \sum_{r=0}^{\infty} \binom{-n-1}{r} \omega^{-r} D^r \qquad (\omega \neq 0)$$

to find an estimate of P_k in (27). Denote by M the maximum of the $\frac{m(m-1)}{2}$ numbers $\frac{1}{|\rho_k - \rho_l|}$ $(1 \leq k < l \leq m)$. Then

$$|P_k(x)| \leq (M^{-1} - D)^{n_k - N} \frac{x^{n_k}}{n_k!} \qquad (x > 0),$$

since the right-hand side, as a polynomial in x, majorizes $P_k(x)$. Furthermore,

$$(M^{-1}-D)^{n_k-N} \frac{x^{n_k}}{n_k!} = M^{N-n_k} \sum_{r=0}^{n_k} \binom{N-n_k+r-1}{r} M^r D^r \frac{x^{n_k}}{n_k!}$$

$$\leq \sum_{r=0}^{n_k} \binom{N}{r} M^{N-n_k+r} x^{n_k-r} \leq (1+1)^N (M+x)^N = (2M+2x)^N \qquad (x > 0),$$

so that

$$(32) \qquad |P_k(1)| \leq (2M+2)^N \qquad (k=1, \ldots, m).$$

§9. ESTIMATION OF $P_k(1)$

If $\rho_1, \ldots, \rho_m$ are algebraic numbers, then also $P_k(1)$ is algebraic. To find an estimate of the denominator of $P_k(1)$ we choose a positive rational integer q such that the $\frac{m(m-1)}{2}$ algebraic numbers $\frac{q}{\rho_k - \rho_l}$ $(1 \leq k < l \leq m)$ all are integral. It follows from (27) and (31), that the number $n_k!\, q^N P_k(1)$ is integral.

§10. The transcendency of e^a for real algebraic $a \neq 0$

Let $a \neq 0$ be a real algebraic number and suppose that also e^a were algebraic. We introduce the algebraic number field K generated by a and e^a, and we denote by h its degree over the rational number field. If ξ is any number in K, the norm $\mathfrak{N}(\xi) = \xi^{(1)} \ldots \xi^{(h)}$ means the product of all h conjugates $\xi^{(1)}, \ldots, \xi^{(h)}$ of ξ, and we define

$$\overline{|\xi|} = \max \quad (|\xi^{(1)}|, \ \ldots, \ |\xi^{(h)}|).$$

Take $m=h+1$, $\rho_k=(k-1)a \quad (k=1,\ldots,m)$, $\quad n_1=n_2=\ldots=n_m=n \geq 1$. The number n will be arbitrarily large, and we shall denote by $c_1, c_2, \ldots$ positive rational integers which are independent of n.

Now the number $P_k(1)$ lies in K. Using (27) for the h conjugate fields we obtain from (32) that

$$\overline{|P_k(1)|} < c_1^{\,n}, \tag{33}$$

since $N=m(n+1)-1=mn+h<c_2 n$. Also

$$R(1) = P_1(1)\, e^{\rho_1} + \ldots + P_m(1)\, e^{\rho_m}$$

lies in K. In virtue of our former estimate for the denominator of $P_k(1)$ we can determine a number

(34) $$T = c_3^n\, n!$$

such that

$$\xi = TR(1)$$

is an <u>integer</u> in K. By (30), this integer is not 0; therefore

(35) $$|\mathfrak{N}(\xi)| \geq 1.$$

On the other hand, by (29) and (34), we have

$$|\xi| < c_4^n (n!)^{-h}$$

for <u>one</u> conjugate, $\xi = \xi^{(1)}$ say, but not necessarily for any other conjugate $\xi^{(2)}, \ldots, \xi^{(h)}$, since the h-1 numbers $e^x (x = \rho_k^{(2)}, \ldots, \rho_k^{(h)})$ are perhaps not conjugate to e^{ρ_k}. However, by (33) and (34), we obtain for <u>all</u> conjugates of ξ the estimate

$$\overline{|\xi|} < c_5^n\, n!\ .$$

Hence

$$|\mathfrak{N}(\xi)| < c_4^n (n!)^{-h} (c_5^n n!)^{h-1} = \frac{c_6^n}{n!} < 1,$$

for all sufficiently large n, contradictory to (35).

This proves that e^a is a transcendental number for all real algebraic $a \neq 0$. In particular, the number e itself is transcendental.

§11. <u>The determinant of m approximation forms</u>

The essential point in the preceding proof was the fact that the algebraic integer ξ is not 0, and this followed from (30). If $\rho_1, \ldots, \rho_m$ again are arbitrary different complex numbers, we can no longer assert that $R(1) \neq 0$. To overcome this difficulty we proceed in the following way.

For any fixed $k = 1, 2, \ldots, m$ we choose $n_l = n \geq 1$ $(l = 1, \ldots, k)$ and $n_l = n-1$ $(l = k+1, \ldots, m)$ and denote the corresponding approximation form by

$$R_k = P_{k1} e^{\rho_1 x} + \ldots + P_{km} e^{\rho_m x} \qquad (k = 1, \ldots, m).$$

The degree of the polynomial P_{kl} is n_l, and this number equals n or $n-1$, according to $k \geq l$ or $k < l$. Now consider the determinant $\Delta = \Delta(x)$ of the P_{kl}. Its $m!$ terms are polynomials in x, and all of them have a degree $< mn$, except the term corresponding to the main diagonal which has the degree mn. It follows that the polynomial Δ has the exact degree mn.

Denoting the minors of the elements of the first column of Δ by $\Delta_1, \ldots, \Delta_m$, we have

$$(36) \qquad \Delta = (\Delta_1 R_1 + \ldots + \Delta_m R_m) e^{-\rho_1 x}$$

The function R_k vanishes at $x = 0$ of order

$$(n_1 + \ldots + n_m) + m - 1 = mn + k - 1 \geq mn;$$

therefore Δ vanishes at $x = 0$ at least of order mn. This proves that

$$\Delta(x) = \gamma x^{mn}, \qquad \Delta(1) = \gamma \neq 0.$$

Now (36) implies that at least one of the m numbers $R_k(1)$ $(k = 1, \ldots, m)$ is not 0, and this suffices, in particular, for the extension of the proof of §10 to the case of a complex algebraic $a \neq 0$.

§12. Algebraic independence

Suppose that p algebraic numbers $a_1, \dots, a_p$ are related by a homogeneous linear equation $g_1 a_1 + \dots + g_p a_p = 0$ with rational integral coefficients $g_1, \dots, g_p$, not all 0. Then the p numbers $\eta_k = e^{a_k}$ $(k=1,\dots,p)$ satisfy the algebraic equation $\eta_1^{g_1} \dots \eta_p^{g_p} = 1$ with rational coefficients. We are going to prove the converse: If $a_1, \dots, a_p$ are algebraic numbers such that $g_1 a_1 + \dots + g_p a_p \neq 0$ for all rational integers $g_1, \dots, g_p$ except for $g_1 = 0, \dots, g_p = 0$, then the p numbers e^{a_k} are not related by an algebraic equation with algebraic coefficients. This means in case $p=1$ that e^a is transcendental for algebraic $a \neq 0$. In particular, π is transcendental, since $e^{2\pi i} = 1$.

Assume that the statement is not true; then there exists a polynomial $G = G(y_1, \dots, y_p)$ in p variables $y_1, \dots, y_p$ and with integral algebraic coefficients, not all 0, such that G vanishes for $y_k = e^{a_k}$ $(k=1,\dots,p)$. Let d be the total degree of $G(y_1, \dots, y_p)$ and denote by h the degree of the algebraic number field K generated by $a_1, \dots, a_p$ and the coefficients of G. We choose a rational integer $f > d$ such that

$$\prod_{k=1}^{p} (f-d+k) > \left(1 - \frac{1}{h}\right) \prod_{k=1}^{p} (f+k); \tag{37}$$

this is possible because the difference of these two polynomials in f has a positive highest coefficient, namely $\frac{1}{h}$.

The number of all monomials $y_1^{g_1} \dots y_p^{g_p}$ of total degree $\leq f$ or $\leq f-d$ is $m = \binom{f+p}{p}$ or $r = \binom{f-d+p}{p}$, respectively. Denoting them by $Y_1, \dots, Y_m$ and $Z_{m-r+1}, \dots, Z_m$, we obtain

$$Z_k G = \alpha_{k1} Y_1 + \dots + \alpha_{km} Y_m \qquad (k=m-r+1,\dots,m),$$

where α_{kl} is 0 or a coefficient of G. It is clear that the r-rowed coefficient matrix (α_{kl}) has the rank r, since the polynomials $Z_k G$ are not related by any homogeneous linear equation with constant coefficients, not all 0.

Let

$$Y_l = y_1^{g_1} \dots y_p^{g_p}, \quad \rho_l = g_1 a_1 + \dots + g_p a_p$$
$$(l=1,\dots,m).$$

Then $\rho_1,\dots,\rho_m$ are m <u>different</u> numbers in K and

$$\alpha_{k1} e^{\rho_1} + \dots + \alpha_{km} e^{\rho_m} = 0 \qquad (k=m-r+1,\dots,m).$$

Now consider the m approximation forms

$$R_k(x) = P_{k1}(x)\, e^{\rho_1 x} + \dots + P_{km}(x)\, e^{\rho_m x}$$
$$(k=1,\dots,m)$$

of §11. We know that the m-rowed matrix $(P_{kl}(1)$ has a non-vanishing determinant $\Delta(1)$, so we can find m-r rows, say for $k=k_1,\dots,k_{m-r}$, which together with the r rows of the matrix (α_{kl}) constitute a non-vanishing determinant. Write

$$P_{k_t l}(1) = \alpha_{tl}, \qquad R_{k_t}(1) = \beta_t \qquad (t=1,\dots,m-r);$$

then

$$\alpha_{k1} e^{\rho_1} + \dots + \alpha_{km} e^{\rho_m} = \beta_k \qquad (k=1,\dots,m)$$

where $\beta_k=0$ for $k=m-r+1,\dots,m$. Let A be the determinant of the α_{kl} and $A_1,\dots,A_m$ the minors corresponding to the first column. It follows that

$$(38) \qquad 0 \neq \mathbf{A} = (\mathbf{A}_1\beta_1 + \ldots + \mathbf{A}_{m-r}\beta_{m-r})\, e^{-\rho_1}.$$

Now we shall apply the results of §9. Since $n_1 = n$ or $n-1$ we can determine a number

$$T = c_7{}^n n!$$

such that the $(m-r)m$ numbers $T\alpha_{kl}$ $(k=1,\ldots,m-r; l=1,\ldots,m)$ are <u>integers</u> in K. This implies that also $T^{m-r}\mathbf{A}$ is integral, whence

$$(39) \qquad |\mathfrak{N}(\mathbf{A})| \geq T^{h(r-m)} = c_8{}^{-n}(n!)^{h(r-m)}.$$

Moreover, by (32),

$$(40) \qquad \overline{|\alpha_{kl}|} < c_9^n \qquad (k=1,\ldots,m;\ l=1,\ldots,m)$$

$$(41) \qquad \overline{|\mathbf{A}|} < c_{10}{}^n.$$

On the other hand, by (29),

$$|\beta_t| < c_{11}n^m(n!)^{-m} < c_{12}{}^n(n!)^{-m}$$

$$(t=1,\ldots,m-r),$$

so that, by (38) and (40),

$$(42) \qquad |\mathbf{A}| < c_{13}{}^n(n!)^{-m}.$$

The estimates (41) and (42) imply

$$|\mathfrak{N}(\mathbf{A})| < c_{14}{}^n(n!)^{-m}.$$

Letting $n \longrightarrow \infty$ and comparing with (39) we see that

$$m \leq h(m-r)$$

$$r \leq (1-\frac{1}{h})m.$$

Because of the definitions of m and r, this contradicts (37).

The result can be formulated in another way. Suppose that $b_1,\ldots,b_r$ are <u>different</u> algebraic numbers. If p of these r numbers, and not more, are linearly independent in the field of rational numbers, then we can find p linearly independent algebraic numbers $a_1,\ldots,a_p$, such that

$$b_k = g_{k1}a_1 + \ldots + g_{kp}a_p \qquad (k=1,\ldots,r)$$

with rational integral coefficients g_{kl}. Now consider the rational function

$$f(y_1,\ldots, y_p) = \sum_{k=1}^{r} c_k y_1^{g_{k1}} \ldots y_p^{g_{kp}}$$

of the variables $y_1,\ldots,y_p$ with arbitrary algebraic coefficients $c_1,\ldots,c_r$, not all 0. It cannot vanish identically in $y_1,\ldots,y_p$, since the sequences of exponents $g_{k1},\ldots,g_{kp}$ are different, for $k=1,\ldots,r$. Now our result shows that f cannot vanish for $y_1=e^{a_1},\ldots,y_p=e^{a_p}$; hence

$$c_1e^{b_1} + \ldots + c_re^{b_r} \neq 0. \tag{43}$$

This is the Lindemann-Weierstrass theorem: If $b_1,\ldots,b_r$ are different algebraic numbers, then $e^{b_1},\ldots,e^{b_r}$ are not related by a homogeneous linear equation with algebraic coefficients, not all 0.

Vice versa, this theorem contains our previous result, since an algebraic equation between $e^{a_1},\dots,e^{a_p}$ with algebraic coefficients would mean that a certain polynomial $f(y_1,\dots,y_p)$ vanishes for $y_1=e^{a_1},\dots,y_p=e^{a_p}$. Since $a_1,\dots,a_p$ are linearly independent in the rational number field, we would obtain a contradiction to (43).

§13. Another expression for the remainder term R(x)

We shall devote the rest of this chapter to study more closely the analytical properties of approximation forms. We start from the complex integral

$$(44). \qquad J = \frac{1}{2\pi i}\int_C \frac{e^{xz}}{Q(z)}\,dz, \qquad Q(z) = \prod_{k=1}^{m} (z-\rho_k)^{n_k+1},$$

where the ρ_k and n_k have their former meaning and C is a simply closed curve in positive direction which contains $\rho_1,\dots,\rho_m$ in its interior. Inserting

$$e^{xz} = e^{x\rho_k}\sum_{l=0}^{\infty}\frac{x^l(z-\rho_k)^l}{l!}$$

we see that the residue of the integrand at $z=\rho_k$ takes the form $Q_k e^{\rho_k x}$, where $Q_k=Q_k(x)$ is a polynomial in x of degree $\leq n_k$. Then, by the residue theorem,

$$J = Q_1e^{\rho_1x} + \dots + Q_me^{\rho_mx}.$$

On the other hand,

$$J = \sum_{l=0}^{\infty} a_l\frac{x^l}{l!}, \qquad a_l = \frac{1}{2\pi i}\int_C \frac{z^l}{Q(z)}\,dz.$$

If we take for C a curve which contains the whole circles $|z| \leq \rho_k$ $(k=1,\ldots,m)$ in its interior, we may use the descending power series expansion

$$\frac{1}{Q(z)} = z^{-N-1} \prod_{k=1}^{m} \left(1-\frac{\rho_k}{z}\right)^{-n_k-1} = z^{-N-1} + \ldots .$$

This proves that

$$a_l = 0 \quad (l=0,1,\ldots,N-1), \quad a_N=1.$$

Now the uniqueness theorem of §7 shows that $Q_k=P_k(x)$ and $J=R(x)$.

The expression

$$R(x) = \frac{1}{2\pi i} \int_C \frac{e^{xz}}{Q(z)}\, dz \tag{45}$$

of the remainder term as a simple complex integral is more elegant than the expression as an $(m-1)$-fold real integral in (28), but it is inconvenient if one wants to prove the results of §8. Without using the uniqueness theorem we can transform (45) into (28) in the following way. Suppose $x>0$, then C can be deformed into a straight line L from $c-i\infty$ to $c+i\infty$, where c is a real number greater than the real parts of the ρ_k. Substitute

$$(z-\rho_k)^{-n_k-1} = \frac{1}{n_k!} \int_0^\infty t_k^{n_k} e^{(\rho_k-z)t_k} dt_k \qquad (k=1,\ldots,m-1)$$

and interchange the order of integration; then

$$R(x) = \int_0^\infty \cdots \int_0^\infty \left\{ \frac{1}{2\pi i} \int_L \frac{e^{(x-t_1-\cdots-t_{m-1})z}}{(z-\rho_m)^{n_m+1}} dz \right\} \prod_{k=1}^{m-1} \frac{t_k^{n_k}}{n_k!} e^{\rho_k t_k} dt_k$$

But

$$\frac{1}{2\pi i} \int_L \frac{e^{tz}}{(z-\rho_m)^{n+1}} dz = \begin{cases} \frac{t^n}{n!} e^{\rho_m t} & (t>0) \\ 0 & (t<0), \end{cases}$$

and (28) follows immediately.

It is simple to get rid of the restriction $x>0$ in (28): Substitute $t_k x$ for t_k, then

$$(46) \qquad R(x) = x^N \underset{\substack{t_1 + \cdots + t_m = 1 \\ t_1 > 0, \ldots, t_m > 0}}{\int \cdots \int} \prod_{k=1}^{m} \left(\frac{t_k^{n_k}}{n_k!} e^{\rho_k t_k x} \right) dt_1 \ldots dt_{m-1},$$

and this holds for arbitrary complex x. Since $R(x) = \frac{x^N}{N!} + \ldots$, the integral in (46) has for $x=0$ the value $\frac{1}{N!}$; hence

$$|R(x)| \le \frac{|x|^N}{N!} e^{\rho|x|}, \qquad \rho = \max(|\rho_1|, \ldots, |\rho_m|),$$

which is a refinement of (29).

It remains to prove formula (27) for P_k from our present point of view. We know that $P_k(x)$ is the coefficient of $(z-\rho_k)^{n_k}$ in the power series expansion of the function

$$f_k(z) = e^{x(z-\rho_k)} \prod_{\substack{l=1 \\ l \neq k}}^{m} (z-\rho_l)^{-n_l-1}$$

at the point $z=\rho_k$,

$$P_k(x) = \frac{1}{n_k!} \{D_z^{n_k} f_k(z)\}_{z=\rho_k} .$$

Putting

$$\prod_{\substack{l=1 \\ l \neq k}}^{m} (z-\rho_l)^{-n_l-1} = g(z),$$

we obtain

$$D_z^{n_k} f_k(z) = e^{x(z-\rho_k)} (x+D_z)^{n_k} g(z).$$

Now consider more generally the expression $\varphi(x+D_z)g(z)$, for any polynomial $\varphi(x)$. Then, by Taylor's formula,

$$\varphi(x+D_z)g(z) = \sum_{l=0}^{\infty} \frac{D_x^{\,l}\varphi(x) D_z^{\,l} g(z)}{l!} = g(z+D_x)f(x).$$

It follows that

$$P_k(x) = g(\rho_k+D)\frac{x^{n_k}}{n_k!} = \prod_{\substack{l=1 \\ l \neq k}}^{m} (\rho_k-\rho_l+D)^{-n_l-1} \frac{x^{n_k}}{n_k!} ;$$

q.e.d.

§14. The interpolation formula

The integral J of §13 also appears in the solution of the following interpolation problem: Let an analytic function $f(z)$ be regular in a domain D of the complex z-plane, and let n points $z_1, \dots, z_n$ of D be given; to determine a polynomial H_{n-1} of degree $\leq n-1$ such that the fraction

$$T_n(z) = \frac{f(z)-H_{n-1}(z)}{\prod_{k=1}^{n}(z-z_k)}$$

is regular in D.

It is clear that there cannot be more than one solution, since the difference of two solutions H_{n-1} would be a polynomial of degree $< n$ and divisible by the polynomial $\prod_{k=1}^{n}(z-z_k)$ of degree n.

Put

$$F_k(z) = (z-z_1) \dots (z-z_k) \qquad (k=0,\dots,n),$$

so that

$$F_k(z) = (z-z_k)F_{k-1}(z) \qquad (k=1,..,n)$$

$$(z-z_k)F_{k-1}(\zeta) - F_k(\zeta) = (z-\zeta)F_{k-1}(\zeta)$$

(47) $$\frac{1}{z-\zeta}\left(\frac{F_{k-1}(\zeta)}{F_{k-1}(z)} - \frac{F_k(\zeta)}{F_k(z)}\right) = \frac{F_{k-1}(\zeta)}{F_k(z)} .$$

Suppose that also ζ lies in D and consider a simply closed curve C in D whose interior lies in D and contains the n+1 points $z_1, \dots, z_n, \zeta$. Define

(48) $$a_{k-1} = \frac{1}{2\pi i}\int_C \frac{f(z)}{F_k(z)}\,dz,$$

$$G_k(\zeta) = \frac{1}{2\pi i}\int_C \frac{F_k(\zeta)}{F_k(z)}\,\frac{f(z)}{z-\zeta}\,dz;$$

then

$$G_0(\zeta) = f(\zeta)$$

and, by (47),

$$G_{k-1}(\zeta) - G_k(\zeta) = a_{k-1}F_{k-1}(\zeta) \quad (k=1,\dots,n);$$

whence

$$f(\zeta) = a_0F_0(\zeta) + a_1F_1(\zeta) + \dots + a_{n-1}F_{n-1}(\zeta) + G_n(\zeta).$$

Therefore the solution of the interpolation problem is given by

$$H_{n-1}(\zeta) = a_0F_0(\zeta) + a_1F_1(\zeta) + \dots + a_{n-1}F_{n-1}(\zeta)$$

$$T_n(\zeta) = \frac{G_n(\zeta)}{F_n(\zeta)} = \frac{1}{2\pi i}\int_C \frac{f(z)}{F_n(z)}\,\frac{dz}{z-\zeta}.$$

Now consider an infinite sequence of points $z_1, z_2, \dots$ in D and suppose that $\lim_{n\to\infty} G_n(z)=0$ for all z in a domain $D_0 \subset D$. Then

$$f(z) = a_0F_0(z) + a_1F_1(z) + \dots \quad (z \text{ in } D_0),$$

and it follows that $a_n \neq 0$ for infinitely many n except when f(z) is a polynomial.

We apply this to the special case $f(z)= e^{xz}$ and take for $z_1,\dots,z_{N+1}$ the set consisting of n_k+1 times the point $\rho_k (k=1,\dots,m)$. By (44) and (48), we see that the coefficient a_N is exactly the integral $J=R(x)$ of §13. In particular, choose $n_k=n$ $(k=1,\dots,r+1)$ and $n_k=n-1$ $(k=r+2,\dots,m)$ and take $n=0,1,\dots$; $r=0,1,\dots,m-1$, so that $N=mn+r=0,1,\dots$. It is easy to see from the expression for $G_N(\zeta)$ that $\lim_{N\to\infty} G_N(z) = 0$ for all z. Since e^z is not a polynomial, this implies that $R(1)\neq 0$ for infinitely many N. The finer algebraical approach of §11 showed that for every $n \geq 0$ the interval $mn \leq N < m(n+1)$ contains at least one N with $R(1) \neq 0$; but this fact is not needed for the proof of the transcendency of e^a for algebraic $a \neq 0$, though it is important for the more general problem solved in §12.

§15. Concluding remarks

It should be mentioned that the preceding proofs of the transcendency of e and π and of the algebraic independence of $e^{a_1}, \ldots, e^{a_p}$, for linearly independent algebraic $a_1, \ldots, a_p$, are not the simplest to be found in literature. Our proofs are related to the original work of Hermite; however, our procedure in constructing the approximation forms is somewhat more algebraic, and this has been decisive for the generalization which we shall investigate in the next chapter.

Two characteristic properties of the exponential function $y = e^x$ were used in our proofs, namely the differential equation $y' = y$ and the addition theorem $e^{x+t} = e^x e^t$. Our generalizations will go in two different directions: Either we consider solutions of linear differential equations without assuming an addition theorem, or we deal with functions satisfying an algebraic addition theorem. In the first case we are led to the problems considered in the second chapter. The second case brings us to the study of elliptic functions from the arithmetic point of view, in the last chapter, and of the function a^x for algebraic $a \neq 0$, in the third chapter.

CHAPTER II

SOLUTIONS OF LINEAR DIFFERENTIAL EQUATIONS

The irrationality of $\frac{\operatorname{tg} a}{a}$ for rational $a^2 \neq 0$, which includes the irrationality of π, was discovered by Lambert nearly two hundred years ago. Lambert's work was generalized by Legendre who considered the power series

$$y = f_\alpha(x) = \sum_{n=0}^{\infty} \frac{x^n}{n!\alpha(\alpha+1)\ldots(\alpha+n-1)} \qquad (\alpha \neq 0,-1,-2,\ldots)$$

satisfying the linear differential equation of second order

$$xy'' + \alpha y' = y.$$

He obtained the continued fraction expansion

$$\frac{y}{y'} = \alpha + \cfrac{x}{\alpha+1+\cfrac{x}{\alpha+2+\ddots}}$$

and proved the irrationality of y/y' for all rational $x \neq 0$ and all rational $\alpha \neq 0,-1,-2,\ldots$. In the special case $\alpha = \frac{1}{2}$ we have

$$y = \cosh(2\sqrt{x}), \qquad y' = \sinh(2\sqrt{x})/\sqrt{x},$$

so that Legendre's theorem contains the irrationality of $\operatorname{tg} a/a$ for rational $a^2 \neq 0$. In more recent times,

Stridsberg proved the irrationality of y and of y', separately, for rational $x \neq 0$ and rational $\alpha \neq 0, -1, \ldots$, and Maier showed that neither y nor y' is a quadratic irrationality. Maier's work suggested the idea of introducing more general approximation forms which enabled me to prove that the numbers y and y' are not connected by any algebraic equation with algebraic coefficients, for any algebraic $x \neq 0$ and any rational $\alpha \neq 0, \pm\frac{1}{2}, -1, \pm\frac{3}{2}, \ldots$. The excluded case of an integer $\alpha + \frac{1}{2}$ really is an exception, since then the function $f_\alpha(x)$ satisfies an algebraic differential equation of first order whose coefficients are polynomials in x with rational numerical coefficients; this follows from the explicit formulas

$$f_{k+\frac{1}{2}} = \frac{1}{2} \cdot \frac{3}{2} \cdots (k - \tfrac{1}{2})\, D^k \cosh (2\sqrt{x}),$$

$$f_{\frac{1}{2}-k} = \frac{(-1)^k x^{k+\frac{1}{2}}}{\frac{1}{2}\cdot\frac{3}{2}\cdots(k-\frac{1}{2})} D^{k+1} \sinh(2\sqrt{x})$$

$$(k=0,1,2,\ldots).$$

For instance, in case $\alpha = \frac{1}{2}$, the differential equation is

$$y^2 - xy'^2 = 1.$$

In the excluded case, however, Lindemann's theorem shows that y and y' are both transcendental for any algebraic $x \neq 0$.

We are now going to develop a general method for transcendency proofs involving solutions of linear differential equations and to apply it to the function $y = f_\alpha(x)$.

A function $y = f(x)$ is called of type E, or an E-function, if

$$f(x) = \sum_{n=0}^{\infty} c_n \frac{x^n}{n!}$$

is a power series satisfying the following three conditions:

1) All coefficients c_n belong to the same algebraic number field of finite degree over the rational number field.

2) If ϵ is any positive number, then $\overline{|c_n|} = O(n^{n\epsilon})$, as $n \to \infty$.

3) There exists a sequence $q_0, q_1, \ldots$ of positive rational integers such that $c_n c_k$ is integral for $k = 0, 1, \ldots, n$ and $n = 0, 1, 2, \ldots$, and that $q_n = O(n^{n\epsilon})$.

We shall recall to memory that the symbol $\overline{|c_n|}$ designates the maximum of the absolute values of all conjugates of the algebraic number c_n. The second condition means that $\overline{|c_n|}$, as a function of n, increases less rapidly than any positive power of n^n. This implies, in particular, that y is an entire function of x. The third condition states that the least common rational denominator of $c_1, \ldots, c_n$ increases less rapidly than any positive power of n^n.

Obviously, any polynomial with algebraic coefficients and the exponential function e^x are examples of E-functions.

It is clear that the derivative y' of an E-function again is an E-function. It is also clear that the sum of two E-functions is an E-function. The same is true for the product: If also

$$g(x) = \sum_{n=0}^{\infty} d_n \frac{x^n}{n!}$$

is an E-function and r_n denotes the least positive rational integral common denominator of $d_0, d_1, \ldots, d_n$, then

$$f(x)g(x) = \sum_{n=0}^{\infty} e_n \frac{x^n}{n!},$$

$$e_n = \sum_{k=0}^{n} \binom{n}{k} c_k d_{n-k},$$

so that

$$\overline{|e_n|} \leq (1+1)^n \max_{k \leq n} \overline{|c_k d_{n-k}|} = 2^n O(n^{2n\epsilon}) = O(n^{3n\epsilon}),$$

and the positive rational integer

$$q_n r_n = O(n^{2n\epsilon})$$

is common denominator of $e_0, \ldots, e_n$. This shows that the E-functions constitute a ring. Finally, also $f(ax)$ is of type E, for any algebraic constant a.

Later we shall study E-functions $y_1 = E_1, \ldots, y_m = E_m$ which satisfy a system of homogeneous linear differential equations of the first order

$$(49) \qquad y'_k = \sum_{l=1}^{m} Q_{kl}(x)\, y_l \qquad (k=1, \ldots, m)$$

whose coefficients Q_{kl} are rational functions of x. Let the polynomial $T(x)$ be the least common denominator of the m^2 rational functions $Q_{kl}(x)$ and consider the numerical coefficients of $T(x)$ and the $T(x)\, Q_{kl}(x)$ as unknown quantities. If we insert the power series $E_1, \ldots, E_m$, then the differential equations (49) become a system of countably many homogeneous linear equations for these finitely many unknown quantities, with algebraic coefficients. Therefore we may assume that the numerical coefficients of the Q_{kl} are integers in K, the algebraic number field generated by all coefficients of $E_1,\ldots,E_m$ together.

It is not hard to show that for power series solutions of (49) the first condition concerning functions of type E, namely, that the coefficients c_n belong to the same algebraic number field K of finite degree, could be weakened to the condition that all coefficients of $E_1,\ldots,E_m$ are algebraic numbers. It could be proved, as a consequence of (49), that the field K generated by all coefficients is of finite degree. However, we do not need this property, and we omit the proof.

It is well known that the system (49) possesses a basis of m solutions $y_k=E_{kl}$ $(k=1,..,m)$, for $l=1,\ldots,m$. Of course, not all E_{kl} are in general of type E; but all E_{kl} are regular at all finite complex points $x=a$ which are different from the zeros of the polynomial $T(x)$, and the determinant of the E_{kl} does not vanish at $x=a$. Any solution of (49) takes the form $y_k=c_1E_{k1}+\ldots+c_mE_{km}$ $(k=1,\ldots,m)$ with constant $c_1,\ldots,c_m$.

§2. Arithmetical lemmas

Consider p homogeneous linear equations with q unknown quantities $x_1,\ldots,x_q$ and integral rational coefficients. They have a non-trivial integral rational solution, if $p<q$. It is useful, for different purposes, to obtain an upper estimate of the absolute values of $x_1,\ldots,x_q$ in terms of bounds for the coefficients.

LEMMA 1. Let

$$y_k = a_{k1}x_1 + \ldots + a_{kq}x_q \qquad (k=1,\ldots,p) \tag{50}$$

be p linear forms with integral rational coefficients and q variables, where $0<p<q$, and suppose that the absolute values of all a_{kl} are not greater than a given positive rational integer A; then there exists a non-trivial integral rational solution $x_1,\dots,x_q$ of $y_1=0,\dots,y_p=0$ satisfying the conditions

$$|x_k| < 1 + (qA)^{\frac{p}{q-p}} \qquad (k=1,\dots,q).$$

PROOF: Let H be a positive rational integer and insert in (50) for $x_1,\dots,x_q$ independently the $2H+1$ values $0, \pm 1, \dots, \pm H$. We obtain $(2H+1)^q$ points with integral coordinates $y_1,\dots,y_p$, all lying in the cube

$$-qAH \leq y_k \leq qAH \qquad (k=1,\dots,p).$$

Since there are exactly $(2qAH+1)^p$ different points with integral coordinates in this cube, it follows that at least two different systems $x_1,\dots,x_q$ have the same image point $y_1,\dots,y_p$, if

$$(51) \qquad (2qAH+1)^p < (2H+1)^q.$$

Under this condition we obtain by subtraction a non-trivial integral rational solution $x_1,\dots,x_q$ of $y_1=0,\dots,y_p=0$ satisfying the inequalities

$$(52) \qquad |x_k| \leq 2H \qquad (k=1,\dots q).$$

Now choose for 2H the even number in the interval

$$(qA)^{\frac{p}{q-p}} - 1 \leq 2H < (qA)^{\frac{p}{q-p}} + 1$$

of length 2; then (51) is fulfilled, because of

$$(2qAH+1)^p < (qA)^p(2H+1)^p$$

$$\leq (2H+1)^{q-p}(2H+1)^p = (2H+1)^q,$$

and the lemma follows from (52).

Now we generalize lemma 1 to algebraic number fields K.

LEMMA 2: Suppose that the coefficients of the p linear forms $y_k = a_{k1}x_1 + \ldots + a_{kq}x_q$ $(k=1,\ldots,p;\ p<q)$ are integers in K, and let $\overline{|a_{kl}|} \leq A$; then there exists in K a non-trivial integral solution $x_1,\ldots,x_q$ of $y_1=0,\ldots,y_p=0$ such that

$$(53) \qquad \overline{|x_k|} < c + c(cqA)^{\frac{p}{q-p}} \qquad (k=1,\ldots,q),$$

where c is a positive constant which only depends upon K.

PROOF: Choose a basis $b_1,\ldots,b_h$ of the integers in K relative to the rational number field. Then any integer a in K has the form $a = g_1b_1 + \ldots + g_hb_h$ with rational integral $g_1,\ldots,g_h$. Solving for $g_1,\ldots,g_h$ from the h equations for the conjugates of a, it follows that $|g_k| < \gamma_1 \overline{|a|}$, where γ_1 only depends upon the choice of the basis. Now write $x_k = x_{k1}b_1 + \ldots + x_{kh}b_h$ with rational integral $x_{k1},\ldots,x_{kh}$ and express also all pqh products $a_{kl}b_r$ in terms of the basis; then the p equations $y_1=0,\ldots,y_p=0$ for $x_1,\ldots,x_q$ become ph homogeneous linear equations for the qh rational integers $x_{11},\ldots,x_{qh}$ with rational integral coefficients of absolute value less than $\gamma_1 \max \overline{|a_{kl}b_r|} < \gamma_2 A$, where γ_2 is a positive rational integer depending upon the basis. Applying lemma 1 we obtain a non-trivial solution satisfying

$$|x_{kl}| < 1 + (\gamma_2 hqA)^{\frac{p}{q-p}}$$

$$(k=1,\ldots,q;\ l=1,\ldots,h),$$

and this implies (53).

§3. Approximation forms

Consider m power series $E_1,\ldots,E_m$ and m polynomials $P_1(x),\ldots,P_m(x)$ of degree $\leq \nu$ with indeterminate coefficients. Obviously, we can choose the $m(\nu+1)$ coefficients, not all 0, such that the approximation form $P_1E_1+\ldots+P_mE_m$ vanishes at x=0 of order $m(\nu+1)-1$ at least. For our further purposes we are interested in the case that $E_1,\ldots,E_m$ are of type E; then the coefficients of $P_1,\ldots,P_m$ can be taken as integers in the algebraic number field K generated by the coefficients of $E_1,\ldots,E_m$. Lemma 2 gives us an upper estimate of the absolute values of the coefficients of $P_1,\ldots,P_m$ and their conjugates; however, the estimate in Lemma 2 is useful only in case the ratio p/q is not too near to its upper bound 1. Therefore we now weaken the condition that the approximation form vanishes at x=0 of the highest possible order.

<u>LEMMA</u> 3: Let $E_1,\ldots,E_m$ be E-functions with coefficients in K, and let an integer n=1,2,... be given. There exist m polynomials $P_1(x),\ldots,P_m(x)$ of degree $\leq 2n-1$ with the following three properties:

1) The coefficients of $P_1,\ldots,P_m$ are integers in K, not all 0, and the maximum of the absolute values of all their conjugates is $O(n^{(2+\epsilon)n})$, for every given positive ϵ and $n\to\infty$.

2) The approximation form

$$(54) \qquad R = P_1E_1 + \dots + P_mE_m = \sum_{\nu=0}^{\infty} a_\nu \frac{x^\nu}{\nu!}$$

vanishes at x=0 of order (2m-1)n at least, so that

$$(55) \qquad a_\nu = 0 \qquad (\nu = 0, 1, \dots, 2mn-n-1).$$

3) The coefficients a_ν of R satisfy the condition

$$a_\nu = \nu^{\epsilon\nu} \quad O(n^{2n}) \qquad (\nu \geq 2mn-n)$$

uniformly in ν.

PROOF: Put

$$E_k = \sum_{\nu=0}^{\infty} c_{k\nu} \frac{x^\nu}{\nu!} \qquad (k=1, \dots, m)$$

and

$$(56) \qquad P_k = (2n-1)! \sum_{\nu=0}^{2n-1} g_{k\nu} \frac{x^\nu}{\nu!}$$

with integral $g_{k\nu}$ in K, so that P_k is a polynomial of degree $\leq 2n-1$ with integral coefficients in K. Computing the coefficients a_ν in the expression (54) we find

$$(57) \qquad P_kE_k = (2n-1)! \sum_{\nu=0}^{\infty} d_{k\nu} \frac{x^\nu}{\nu!},$$

$$d_{k\nu} = \sum_{\rho=0}^{2n-1} \binom{\nu}{\rho} g_{k\rho} c_{k,\nu-\rho}$$

$$(58) \qquad a_\nu = (2n-1)!(d_{1\nu} + \dots + d_{m\nu}) \qquad (\nu = 0, 1, \dots).$$

The condition (55) yields $(2m-1)n$ homogeneous linear equations for the $2mn$ unknown quantities $g_{k\rho}$ ($k=1,\ldots,m$; $\rho=0,\ldots,2n-1$). Determine a positive rational integer q_n such that $q_n c_{k\nu}$ is integral for $k=1,\ldots,m$ and $\nu=0,\ldots,2mn-n-1$; because of condition 3) in the definition of a function of type E, we may choose $q_n=O(n^{\epsilon n})$. Multiply the equations $a_\nu=0$ by $q_n/(2n-1)!$; then the coefficient $q_n\binom{\nu}{\rho}c_{k,\nu-\rho}$ of $g_{k\rho}$ lies in K, and all its conjugates again are $O(n^{\epsilon n})$, because of

$$(59) \qquad \binom{\nu}{\rho} \leq 2^\nu, \qquad\qquad \overline{|c_{k\nu}|} = O(\nu^{\epsilon\nu})$$

and $\nu < (2m-1)n$. Applying lemma 2 with $p=(2m-1)n$, $q=2mn$, $A=O(n^{\epsilon n})$, we find integral $g_{k\nu}$ ($k=1,\ldots,m$; $\nu=0,\ldots,2n-1$) in K, not all 0, such that (55) is fulfilled and

$$\overline{|g_{k\nu}|} = O(n^{\epsilon n}),$$

because of $\frac{p}{q-p} = 2m-1=O(1)$. The remaining statements of the lemma readily follow from (56), (57), (58), (59), together with the estimate $(2n-1)! = O(n^{2n})$.

§4. Normal systems

Suppose that the E-functions $E_1,\ldots,E_m$ satisfy a system of homogeneous linear differential equations of first order (49) whose coefficients Q_{kl} are rational functions of x, with integral numerical coefficients in the algebraic number field K generated by the coefficients of $E_1,\ldots,E_m$. It may happen that the matrix $Q=(Q_{kl})$ decomposes into a number r of quadratic boxes $Q_t=(Q_{kl,t})$, where $t=1,\ldots,r$, with $m_1,\ldots,m_r$ rows, so that $k,l=1,\ldots,m_t$ and $m_1+\ldots+m_r=m$. This means that the boxes are arranged in the diagonal of Q and that all elements of Q outside of the boxes are 0. The decomposition is unique if we choose r as large as possible; then we call $Q_1,\dots,Q_r$ the primitive parts of Q. Of course, it can happen that r=1 and Q itself is primitive.

Corresponding to the decomposition of Q into primitive parts the system (49) breaks into r separate systems

$$(60)\qquad y'_{k,t} = \sum_{l=1}^{m_t} Q_{kl,t}(x)y_{l,t} \qquad (k=1,\dots,m_t;t=1,\dots,r).$$

Let $y_{k,t}=y_{kl,t}(k=1,\dots,m_t)$, for $l=1,\dots,m_t$, be a basis for the solutions of (60); then the m-rowed matrix Y consisting of the r boxes $Y_t=(y_{kl,t})$ is a solution matrix for (49).

Now consider any solution $y_1,\dots,y_m$ of (49) and introduce the sum

$$(61)\qquad R = P_1y_1 + \dots + P_my_m$$

whose coefficients $P_1,\dots,P_m$ are arbitrary polynomials in x. We are interested in the case that R vanishes identically in x. Using the box decomposition of Q we write $P^*_{k,t}(k=1,\dots,m_t;\ t=1,\dots,r)$ instead of $P_1,\dots,P_m$. Expressing the solution $y_1,\dots,y_m$ in terms of the basis we then obtain

$$(62)\qquad R = \sum_{k,l,t} P^*_{k,t}c_{l,t}\, y_{kl,t},$$

where $k,l=1,\dots,m_t$ and $t=1,\dots,r$, the $P^*_{k,t}$ being polynomials and the $c_{l,t}$ constant. The sum R trivially equals 0 if all products $P^*_{k,t}c_{l,t}$ are 0, or, in other words, if for each $t=1,\dots,r$ either all polynomials $P^*_{k,t}$ are identically 0 or all constants $c_{l,t}$ are 0. We shall say that the boxes Y_t are <u>independent</u>, if R does not vanish identically in x, except in the trivial case. If the box Q_t is given, then the solution box Y_t only is determined up to a factor C_t to the right, where C_t is an arbitrary m_t-rowed non-singular constant matrix; but the passage from Y_t to Y_tC_t does not affect the independence property.

We shall study the algebraical meaning of independence of the solution boxes. Define $R_1=R$ by (61) and

$$R_{k+1} = T\, R_k' \qquad (k=1,2,\dots) \tag{63}$$

where $T(x)$ is the least common denominator of the $Q_{kl}(x)$. Because of (49) we can write

$$R_k = P_{k1}y_1 + \dots + P_{km}y_m \qquad (k=1,2,\dots), \tag{64}$$

where

$$P_{k+1,l} = T\Big(P_{kl}' + \sum_{g=1}^{m} P_{kg}Q_{gl}\Big) \qquad (l=1,\dots,m) \tag{65}$$

and

$$P_{1l} = P_l. \tag{66}$$

Clearly, all P_{kl} are polynomials in x. Denote by $\Delta = \Delta(x)$ the determinant with the elements P_{kl} $(k,l=1,\dots,m)$ and by Δ_{lk} the minor of P_{kl}; then (64) implies

$$\Delta\, y_k = \sum_{l=1}^{m} \Delta_{kl}R_l \qquad (k=1,\dots,m). \tag{67}$$

There are two cases when we can immediately assure that $\Delta(x)$ vanishes identically in x: If in (62) all $P^*_{k,t}$, for $k=1,\dots,m_t$ and at least one t, are identically 0, then the box decomposition of Q shows, because of (65), that m_t columns of Δ are 0; if R is identically 0, but not all $y_1, \ldots, y_m$, then (63) and (64) imply $\Delta = 0$. Exclude the first case; then the assumption of the second case, because of (62), means that the boxes are not independent. Now we prove that, in all other cases, the polynomial $\Delta(x)$ does not vanish identically.

LEMMA 4: Suppose that the boxes Y_t are independent and that, for each $t=1,\ldots,r$, not all $P^*_{k,t}$ $(k=1,\ldots,m_t)$ vanish identically; then $\Delta(x)$ is not identically 0.

PROOF: If $\Delta = 0$, then we could determine $\mu \leq m$ polynomials $A_1, \ldots, A_\mu$ such that

$$A_1 P_{1l} + \ldots + A_\mu P_{\mu l} = 0 \qquad (l=1,\ldots,m), \qquad A_\mu \neq 0.$$

Let $y_1, \ldots, y_m$ be a completely arbitrary solution of (49) and use the notation (61) and (62). It follows from (63) and (65) that

$$A_1 R_1 + \ldots + A_\mu R_\mu = 0$$

$$(68) \quad B_1 R^{(\mu-1)} + B_2 R^{(\mu-2)} + \ldots + B_\mu R = 0,$$

where $B_1 = A_\mu T^{\mu-1}$ and $B_2, \ldots, B_\mu$ are polynomials in x. Taking for $y_1, \ldots, y_m$ the m basic solutions, we see that each of the m functions

$$R_{l,t} = \sum_{k=1}^{m_t} P^*_{k,t}\, y_{kl,t} \quad (l=1,\ldots,m_t;\ t=1,\ldots,r)$$

satisfies the homogeneous linear differential equation (68) of order $\mu - 1 < m$. Therefore we have a non-trivial homogeneous linear relationship

$$\sum_{l,t} c_{l,t}\, R_{l,t} = 0 \tag{69}$$

with constant $c_{l,t}$. But the r boxes Y_t are independent, and not all products $P^*_{k,t}\, c_{l,t}$ $(k,l=1,\ldots,m_t;\ t=1,\ldots,r)$ are 0, because at least one $P^*_{k,t} \neq 0$, for each t. So (69) is impossible.

We shall say the m functions $E_1,\ldots,E_m$ of type E form a normal system if they all are not identically 0 and satisfy the m differential equations (49), with rational $Q_{kl}(x)$ and algebraic numerical coefficients, whose solutions boxes Y_t are independent. This condition of normality will be decisive in the proof of an extension of the result of §11, Chapter I, which led to the Lindemann-Weierstrass theorem.

§5. The coefficient matrix of the approximation forms

From now on we shall assume that the E-functions $E_1,\ldots,E_m$ form a normal system and that $P_1,\ldots,P_m$ are the polynomials in the approximation form (54). Again the polynomials P_{kl} $(k=1,2,\ldots;\ l=1,\ldots,m)$ will be defined as in (64) and (66), so that the approximation forms

$$R_k = P_{k1}\, E_1 + \ldots + P_{km} E_m \qquad (k=1,2,\ldots)$$

satisfy the equations

$$R_1 = R, \qquad R_{k+1} = TR'_k\ . \tag{70}$$

None of the functions $E_1,\ldots,E_m$ is identically 0. Suppose that the derivatives $E_l^{(k)}(x)$, for $k = 1,\ldots,p-1$ and $l=1,\ldots,m$, vanish at $x=0$, but not all $E_l^{(p)}(0)$ $(l=1,\ldots,m)$. Of course, p may be 0. We denote by

q the maximum of the degrees of the m^2+1 polynomials T and $T\,Q_{kl}$ $(k,l=1,\dots,m)$, and we define

$$t = n + p + q\,\frac{m(m-1)}{2} - 1.$$

LEMMA 5: Let α be any complex number different from 0 and the zeros of the polynomial $T(x)$ and suppose that

(71) $$n \geq p + q\,\frac{m(m-1)}{2}\ ;$$

then the matrix

(72) $$(P_{kl}(\alpha))_{\substack{k=1,\dots,m+t \\ l=1,\dots,m}}$$

has the rank m.

PROOF: We use the box decomposition of the system (49) and write again more explicitly $P^*_{k,t}$ $(k=1,\dots,m_t;\ t=1,\dots,r)$ instead of $P_1,\dots,P_m$. Not all $P^*_{k,t}$ are identically 0, because of lemma 3. We may choose the notation such that $P^*_{k,t}=0$ for $k=1,\dots,m_t$ and $t=\rho+1,\dots,r$; whereas for every $t\leq\rho$ at least one $P^*_{k,t}\neq 0$. Put $m_1+\dots+m_\rho=\mu$; then $1\leq\mu\leq m$. We shall first prove that $\mu=m$.

We apply lemma 4 with μ instead of m. The assumptions are satisfied, since $E_1,\dots,E_\mu$ form a normal system, a fortiori. Denoting by $\Delta=\Delta(x)$ the determinant of the P_{kl} $(k,l=1,\dots,\mu)$, we see that Δ does not vanish identically in x. By (67),

(73) $$\Delta\, y_k = \sum_{l=1}^{\mu} \Delta_{kl}\,(P_{l1}y_1+\dots+P_{l\mu}\,y_\mu) \qquad (k=1,\dots,\mu)$$

identically in x and $y_1,\dots,y_\mu$, and in particular

$$\Delta E_k = \sum_{l=1}^{\mu} \Delta_{kl} R_l . \tag{74}$$

Because of lemma 3 the power series R vanishes at $x=0$ of order $(2m-1)n$ at least, so that R_l, in view of (70), vanishes there of order $\geq (2m-1)n-l+1$. Choose k such that $E_k^{(p)}(0) \neq 0$; then (74) implies

$$\Delta(x) = x^{(2m-1)n-\mu+1-p} \Delta_0(x), \tag{75}$$

where $\Delta_0(x)$ again is a polynomial and not identically 0. On the other hand, P_{kl} has a degree $\leq 2n-1+(k-1)q$, by (65) and lemma 3; therefore the μ-rowed determinant Δ has a degree $\leq (2n-1)\mu + q\frac{\mu(\mu-1)}{2}$. If b denotes the degree of Δ_0, then

$$\begin{aligned} 0 \leq b &\leq (2n-1)\mu + q\frac{\mu(\mu-1)}{2} - (2m-1)n + \mu - 1 + p \\ &= -2n(m-\mu) + n + p + q\frac{\mu(\mu-1)}{2} - 1 . \end{aligned} \tag{76}$$

Because of the condition (71), the right-hand side in (76) would be negative in case $m-\mu \geq 1$. This proves that $\mu = m$ and

$$b \leq n+p + q\frac{m(m-1)}{2} - 1 = t . \tag{77}$$

The number α in the statement of the lemma satisfies $\alpha T(\alpha) \neq 0$. If $\Delta(x)$ vanishes at $x=\alpha$ of order a; then, by (75) and (77),

$$0 \leq a \leq t . \tag{78}$$

Consider for a moment the indeterminates $y_1, \ldots, y_m$ in

(73) as <u>arbitrary solutions of</u> (49) and apply a times the operator TD. We then obtain the formula

$$T^a(x)\,\Delta^{(a)}(x)y_k + \sum_{l=0}^{a-1} \Delta^{(l)}(x)L_{kl} \tag{79}$$

$$= \sum_{l=1}^{m+a} M_{kl}(x)\left(P_{l1}(x)y_1+\dots+P_{lm}(x)y_m\right) \qquad (k=1,\dots,m),$$

identically in x and the <u>indeterminates</u> $y_1,\dots,y_m$, where the L_{kl} are linear forms in $y_1,\dots,y_m$ whose coefficients are polynomials in x, and the M_{kl} are polynomials in x. Now insert $x=\alpha$; then

$$T^a(\alpha)\Delta^{(a)}(\alpha) = \beta \neq 0$$

and the left-hand side of (79) reduces to $\beta\, y_k$. It follows that $y_1,\dots,y_m$ can be expressed as linear combinations of the m+a linear forms $P_{l1}(\alpha)\, y_1+\dots+P_{lm}(\alpha)\, y_m$ $(l=1,\dots,m+a)$. In view of (78), this contains the statement of the lemma.

§6. <u>Estimation of R_k and P_{kl}</u>

The coefficients of $E_1,\dots,E_m$, T and TQ_{kl} $(k,l=1,\dots,m)$ lie in some algebraic number field K of finite degree. By (65) and lemma 3, the coefficients of all polynomials P_{kl} $(k=1,2,\dots;\ l=1,\dots,m)$ all lie in K.

<u>LEMMA</u> 6: Let α be a number in K and $k\leq m+t$, then

$$|R_k(\alpha)| = O\left(n^{(3+\epsilon)n - (2m-2)n}\right)$$

$$\overline{|P_{kl}(\alpha)|} = O\left(n^{(3+\epsilon)n}\right) \qquad (l=1,\dots,m).$$

PROOF: If the coefficients of two power series $A=\alpha_0+\alpha_1 x+\dots$ and $B=\beta_0+\beta_1 x+\dots$ satisfy the conditions $|\alpha_k|\leq\beta_k$ for $k=0,1,\dots$, we shall write $A \prec B$. Plainly, $T \prec c(1+x)^q$ and $TQ_{kl} \prec c(1+x)^q$ with some positive constant c. We are going to prove by induction that

$$(80)\qquad R_{k+1} \prec c^k(1+x)^{kq} \prod_{\nu=0}^{k-1} (\nu q+D)\,\hat{R} \qquad (k=0,1,\dots)$$

and

$$(81)\qquad P_{k+1,l} \prec c^k(1+x)^{kq+2n-1} \prod_{\nu=0}^{k-1} (\nu q+m+2n-1)\,O(n^{(2+\epsilon)n})$$

$$(l=1,\dots,m),$$

where

$$(82)\qquad \hat{R} = \sum_{\nu=0}^{\infty} |a_\nu| \frac{x^\nu}{\nu!} = \sum_{\nu=(2m-1)n}^{\infty} \nu^{\epsilon\nu} \frac{x^\nu}{\nu!}\, O(n^{2n}),$$

and the estimate (81) remains valid if the coefficients of $P_{k+1,l}$ are replaced by their conjugates.

Because of lemma 3, the statement is true for $k=0$. If it is proved for $k-1 \geq 0$ instead of k, we obtain, by (63) and (65),

$$R_{k+1} = TR'_k \prec c(1+x)^q c^{k-1}$$

$$\{(k-1)q(1+x)^{(k-1)q-1}+(1+x)^{(k-1)q}D\} \prod_{\nu=0}^{k-2} (\nu q+D)\,\hat{R}$$

$$\prec c^k(1+x)^{kq} \prod_{\nu=0}^{k-1} (\nu q+D)\,\hat{R},$$

$$P_{k+1,1} \prec c(1+x)^q c^{k-1}$$

$$\prod_{\nu=0}^{k-2}(\nu q+m+2n-1)\ O\,(n^{(2+\epsilon)n})(m+D)(1+x)^{(k-1)q+2n-1}$$

$$\prec c^k\ (1+x)^{kq+2n-1} \prod_{\nu=0}^{k-1}\ (\nu q+m+2n-1)\ O\ (n^{(2+\epsilon)n}),$$

and this is the required result.

If $k \leq m+t=n+O(1)$, then, by (82),

$$\prod_{\nu=0}^{k-1}\ (\nu q+D)\hat{R} \prec O(n^{(1+\epsilon)n})(1+D)^k\hat{R}$$

$$= O(n^{(3+\epsilon)n}) \sum_{\rho=0}^{k} \sum_{\nu=(2m-1)n}^{\infty} \binom{k}{\rho}\nu^{\epsilon\nu}\ \frac{x^{\nu-\rho}}{(\nu-\rho)!}$$

$$\prec O\ (n^{(3+\epsilon)n})\ 2^k \sum_{\nu=(2m-1)n-k}^{\infty} (\nu+k)^{\epsilon(\nu+k)}\ \frac{x^\nu}{\nu!}$$

$$\prec O\ (n^{(3+\epsilon)n})\ 2^k \sum_{\nu=(2m-1)n-k}^{\infty} \{(2\nu)^{2\epsilon\nu} + (2k)^{2\epsilon k}\}\ \frac{x^\nu}{\nu!}$$

$$(83)\quad \left(\prod_{\nu=0}^{k-1}\ (\nu q+D)\hat{R}\right)_{x=\alpha} = O\ \left(n^{(3+\epsilon)n}\right)\ O\ \left(n^{\epsilon n-(2m-2)n}\right).$$

The lemma now follows from (80), (81), and (83).

§7. The rank of $E_1(\alpha),\ldots,E_m(\alpha)$

Let $\omega_1,\ldots,\omega_m$ be any complex numbers. We shall say that they have rank r relative to the algebraic number field K if they satisfy m-r, and not more than m-r, linearly independent homogeneous linear equations

$$\lambda_{k1}\ \omega_1+\ldots+\lambda_{km}\omega_m = 0 \qquad (k=1,\ldots,m-r)$$

with coefficients λ_{kl} in K. In other words, the set $\omega_1, \ldots, \omega_m$ contains r and not more than r numbers which are linearly independent in K.

LEMMA 7: Suppose that α and all coefficients of $E_1, \ldots, E_m$ lie in the algebraic number field K of degree h. If $E_1, \ldots, E_m$ form a normal system and if $\alpha T(\alpha) \neq 0$, then the rank of $E_1(\alpha), \ldots, E_m(\alpha)$ relative to K is at least $\frac{m}{2h}$.

PROOF: Since α is a regular point of the coefficients Q_{kl} in (49), it follows that not all $E_k(\alpha)=0$ $(k=1, \ldots, m)$. Suppose that

$$(84) \qquad \lambda_{k1}E_1(\alpha) + \ldots + \lambda_{km}E_m(\alpha) = 0 \qquad (k=1, \ldots, m-r)$$

are m-r linearly independent equations for $E_1(\alpha), \ldots, E_m(\alpha)$ with integral coefficients λ_{kl} in K. Since not all $E_k(\alpha)=0$, we have $r \geq 1$.

Now apply lemma 5, and determine r rows of the matrix (72), say for $k=k_1, \ldots, k_r$, such that the r linear forms

$$(85) \qquad P_{k1}(\alpha)\, E_1(\alpha) + \ldots + P_{km}(\alpha)\, E_m(\alpha) = R_k(\alpha) \qquad (k=k_1, \ldots, k_r)$$

together with the m-r linear forms in (84) are independent. Write (85) before (84), denote by Λ the determinant of the m^2 coefficients P_{kl} and λ_{kl}, and by Λ_{kl} the minor of the element in the kth row and l^{th} column of Λ; then

$$\Lambda\, E_l(\alpha) = \sum_{g=1}^{r} \Lambda_{gl}\, R_{k_g}(\alpha) \qquad (l=1, \ldots, m).$$

Because of lemma 6,

$$R_{k_g}(\alpha) = O\left(n^{(3+\epsilon)n-(2m-2)n}\right) \qquad (g=1,\ldots,r),$$

$$\overline{|P_{k_g l}(\alpha)|} = O\left(n^{(3+\epsilon)n}\right) \qquad (l=1,\ldots,m);$$

therefore

$$\Lambda_{gl} = O\left(n^{(3+\epsilon)n(r-1)}\right)$$

$$\Lambda E_1(\alpha) = O\left(n^{(3+\epsilon)rn-(2m-2)n}\right)$$

$$\overline{|\Lambda|} = O\left(n^{(3+\epsilon)rn}\right)$$

$$E_1(\alpha)\,\mathfrak{N}(\Lambda) = O\left(n^{(3+\epsilon)rhn-(2m-2)n}\right).$$

Choose a rational positive integer v such that $v\alpha$ is integral; then $g^{2n-1}P_1(\alpha)$ and $g^{2n-1+(k-1)q}P_{kl}(\alpha)$, for $k=1,2,\ldots$ and $l=1,\ldots,m$, are integral. Since $\Lambda \neq 0$, it follows that

$$e^{O(n)}\mathfrak{N}(\Lambda) > 1.$$

Finally, let $n\to\infty$. The number $E_l(\alpha)\neq 0$ for some l; hence

$$3rh \geq 2m-2$$

$$r \geq \frac{2m-2}{3h}$$

Now the assertion follows, in case $m\geq 4$, because of $\frac{2m-2}{3} \geq \frac{m}{2}$. If $m=3$, we have $r\geq \frac{4}{3h}$; and this means for integers r, h the same as $r \geq \frac{3}{2h}$. In the remaining case m=1,2, the assertion is trivial, because of $r \geq 1$.

§8. Algebraic independence

Consider any solution $y_1, \ldots, y_m$ of the system (49). If ν is any given rational positive integer, then the number of power products

$$Y = y_1^{\nu_1} y_2^{\nu_2} \ldots y_m^{\nu_m},$$

with non-negative rational integral exponents $\nu_1, \ldots, \nu_m$ and

$$\nu_1 + \ldots + \nu_m \leq \nu$$

equals

$$\mu = \mu_\nu = \binom{m+\nu}{m}.$$

Because of

$$D \log Y = \sum_{k=1}^{m} \nu_k D \log y_k,$$

the μ functions Y satisfy a system of μ homogeneous linear differential equations of first order, whose coefficients are homogeneous linear functions of the Q_{kl} with rational integral numerical coefficients

<u>THEOREM</u>: Let the E-functions $E_1, \ldots, E_m$ be solutions of a system of m homogeneous linear differential equations of first order, whose coefficients Q_{kl} are rational functions with algebraic numerical coefficients, and suppose that the μ_ν power products $E_1^{\nu_1} \ldots E_m^{\nu_m}$ $(\nu_1 + \ldots + \nu_m \leq \nu)$ form a normal system, for

all $\nu = 1, 2, \ldots$. If α is any algebraic number different from 0 and the poles of the Q_{kl}, then the m numbers $E_1(\alpha), \ldots, E_m(\alpha)$ are not related by an algebraic equation with algebraic coefficients.

PROOF: Let $S(y_1, \ldots, y_m)$ be a polynomial in $y_1, \ldots, y_m$ whose coefficients are algebraic numbers and not all 0. Denote the total degree of S by s, take $\nu \geq s$ and consider the polynomials $y_1^{\nu_1} \ldots y_m^{\nu_m} S$ with

$$\nu_1 + \ldots + \nu_m \leq \nu - s.$$

Their number equals

$$\mu_{\nu-s} = \binom{m-s+\nu}{m}$$

and their total degree is $\leq \nu$. If S has the zero $y_1 = E_1(\alpha), \ldots, y_m = E_m(\alpha)$, then we obtain $\mu_{\nu-s}$ independent homogeneous linear relations between the μ_ν numbers $E_1^{\nu_1}(\alpha) \ldots E_m^{\nu_m}(\alpha)$ $(\nu_1 + \ldots + \nu_m \leq \nu)$, with coefficients in the algebraic number field K generated by the coefficients of $E_1, \ldots, E_m$, Q_{kl}, S and the number α.

Now apply lemma 7, with μ_ν and the power products $E_1^{\nu_1} \ldots E_m^{\nu_m}$ instead of m and $E_1, \ldots, E_m$. Because of the normality assumption we infer that

$$\mu_\nu - \mu_{\nu-s} \geq \frac{\mu_\nu}{2h}, \tag{86}$$

where h is the degree of the field K. But μ_ν and $\mu_{\nu-s}$ are polynomials of degree m in ν starting with the same term $\frac{\nu^m}{m!}$; therefore (86) contains a contradiction as $\nu \to \infty$.

The importance of the theorem lies in the fact that it reduces the arithmetical problem of algebraic independence of the numbers $E_1(\alpha), \ldots, E_m(\alpha)$ to the analytical problem of normality of the power products of the functions $E_1(x),\ldots,E_m(x)$.

The simplest example is $E_k=e^{a_k x}$ $(k=1,\ldots,m)$, where $a_1,\ldots,a_m$ are algebraic numbers and linearly independent in the rational number field. The μ power products then take the form $e^{\rho_k x}$ $(k=1,\ldots,\mu)$ with μ different algebraic numbers ρ_k. The corresponding system (49) now simply is $y'_k=\rho_k y_k$ $(k=1,\ldots,\mu)$, so that all boxes are one-rowed, and normality means that any equation $P_1 e^{\rho_1 x}+\ldots+P_\mu e^{\rho_\mu x}=0$ with polynomials $P_1,\ldots,P_\mu$ implies $P_1=0,\ldots,P_\mu=0$, which is easily proved. This shows that the Lindemann-Weierstrass theorem is contained in our theorem.

§9. Hypergeometric E-functions

In order to obtain more general applications of the theorem, we have to look for E-functions satisfying homogeneous linear differential equations whose coefficients are rational functions. We have as yet no method of finding all such functions, and we only know the following rather specialized way of construction.

Put

$$[\alpha,\nu]=\alpha(\alpha+1)\ldots(\alpha+\nu-1)\qquad(\nu=0,1,\ldots),$$

so that $[\alpha,0]=1$ and $[\alpha,\nu+1]=(\alpha+\nu)[\alpha,\nu]$. Let $a_1,\ldots,a_g$ and $b_1,\ldots,b_m$ be rational numbers, $b_k\neq 0,-1,-2,\ldots(k=1,\ldots,m)$, and $m-g=t>0$. Define

$$c_n=\frac{[a_1,n][a_2,n]\ldots[a_g,n]}{[b_1,n][b_2,n]\ldots[b_m,n]}\qquad(n=0,1,\ldots),$$

$$y=\sum_{n=0}^{\infty}c_n x^{tn},$$

and introduce the operators

$$\Delta_\lambda\ f(x) = D\left(x^{1+\lambda t} f(x)\right),$$

$$A = \Delta_{a_1-a_2} \Delta_{a_2-a_3} \ldots \Delta_{a_{g-1}-a_g} \Delta_{a_g},$$

$$B = \Delta_{b_1-b_2} \Delta_{b_2-b_3} \ldots \Delta_{b_{m-1}-b_m} \Delta_{b_m-1},$$

then

$$Ax^{tn-1} = (a_1+n)(a_2+n) \ldots (a_g+n)t^g x^{t(a_1+n)-1},$$

$$Bx^{tn-1} = (b_1+n-1)(b_2+n-1) \ldots (b_m+n-1)t^m x^{t(b_1+n-1)-1},$$

so that

$$A\ \frac{y}{x} = t^g x^{ta_1-1} \sum_{n=0}^{\infty} (a_1+n)(a_2+n) \ldots (a_g+n)c_n x^{tn}$$

$$B\ \frac{y}{x} = t^m x^{tb_1-1} \sum_{n=-1}^{\infty} (b_1+n)(b_2+n) \ldots (b_m+n)c_{n+1} x^{tn}.$$

Since

$$(a_1+n)\ldots(a_g+n)c_n = (b_1+n)\ldots(b_m+n)c_{n+1} \qquad (n=0,1,\ldots),$$

we obtain

$$x^{1-g}(x^{-tb_1}B - t^t x^{-ta_1}A)\frac{y}{x} = t^m(b_1-1)(b_2-1)\ldots(b_m-1)x^{-m}.$$

The left-hand side takes the form

$$W = y^{(m)} + Q_1 y^{(m-1)} + \dots + Q_{m-1} y' + Q_m y,$$

where $Q_1, \dots, Q_m$ are polynomials in x^{-1} with rational coefficients. If one of the m numbers $b_1, \dots, b_m$ equals 1, then y satisfies the homogeneous linear differential equation $W = 0$ of order m. If $b_k \neq 1$ for all $k=1,\dots,m$, we replace g,m by g+1, m+1 and define $a_{g+1} = b_{m+1} = 1$; this yields a homogeneous linear differential equation $W=0$ of order m+1.

Now we shall prove that y is an E-function. Writing

$$y = \sum_{\nu=0}^{\infty} d_\nu \frac{x^\nu}{\nu !}$$

we have

$$(87) \qquad d_{tn} = (tn)!\, c_n$$

$$= \frac{[a_1,n]\dots[a_g,n][\ 1,\ n\ ]\dots[\ 1,n]}{[b_1,n]\dots[b_g,n][b_{g+1},n]\dots[b_m,n]} \cdot \frac{(tn)!}{(n!)^t} \qquad (n=0,1,\dots)$$

and $d_\nu = 0$ otherwise. It is clear that

$$\frac{[a,n]}{[b,n]} = \prod_{k=1}^{n} \{ (1+\frac{a-1}{k}) \,/\, (1+\frac{b-1}{k}) \} = O(e^{\epsilon n}) \qquad (b \neq 0,-1,-2,\dots),$$

for any given $\epsilon > 0$ and $n \to \infty$, and

$$\frac{(tn)!}{(n!)^t} \leq (1+\dots+1)^{tn} = t^{nt};$$

hence

$$d_n = O(n^{\epsilon n}).$$

It remains to prove that the third condition in the definition of an E-function is fulfilled, namely, that the least common denominator of the rational numbers $d_0, \ldots, d_n$ is $O(n^{\epsilon n})$. Since the number $(tn)!/(n!)^t$ is integral, it suffices to prove, because of (87), that the least common denominator of the n+1 numbers $[a,k] / [b,k]$ $(k=0,\ldots,n)$ is $O(n^{\epsilon n})$, for any rational a, b and $b \neq 0,-1,-2,\ldots$. Put $a=\alpha/\beta$, $b=\gamma/\delta$, where $(\alpha,\beta)=1$, $(\gamma,\delta)=1$, $\delta>0$, then

$$\frac{\beta^{2k}[a,k]}{\delta^k\,[b,k]} = \frac{\beta^k\alpha(\alpha+\beta)(\alpha+2\beta)\ldots(\alpha+(k-1)\beta)}{\gamma(\gamma+\delta)(\gamma+2\delta)\ldots(\gamma+(k-1)\delta)} = \frac{M_k}{N_k} \qquad (k=0,\ldots,n),$$

say. Consider any prime factor p of N_k; then $(p,\delta)=1$. If ν runs over p^l consecutive rational integers, then one and only one of the p^l corresponding integers $\gamma+\nu\delta$ is divisible by p^l, for $l=1,2,\ldots$. Therefore at least $[kp^{-l}]$ and at most $1+[kp^{-l}]$ of the k factors $\gamma, \gamma+\delta, \ldots, \gamma+(k-1)\delta$ in N_k are divisible by p^l. In case $p^l > |\gamma|+(k-1)\delta$ none of these factors is divisible by p^l. Suppose that N_k is divisible by p^s, but not by p^{s+1}, then

$$\sum_l [kp^{-l}] \leq s \leq \sum_l (1+[kp^{-l}]),$$

where $l=1,2,\ldots$ is restricted by the condition

$$l \leq \log(|\gamma| + (k-1)\delta) / \log p.$$

Hence

$$(88)\qquad [\tfrac{k}{p}] \leq s \leq [\tfrac{k}{p}] + O(\frac{k}{p^2}) + O(\frac{\log(k+1)}{\log p})$$

and

$$p \leq |\gamma| + (k-1)\delta \ .$$

In case $(p,\beta)=1$ the left-hand estimate in (88) shows that also the numerator M_k is divisible by $p^{[\frac{k}{p}]}$, and this remains true for the prime factors of β, because of the factor β^k in M_k. Denote by r_p the exponent of p in the reduced denominator of M_k/N_k; then

$$r_p = O(\frac{n}{p^2}) + O(\frac{\log n}{\log p}),$$

for k=0,1,...,n. Therefore the least common denominator q_n of $[a,k]/[b,k]$ $(k=0,\dots,n)$ satisfies

$$\log q_n = O(n) + \sum_{p=O(n)} \{O(\frac{n}{p^2}) + O(\frac{\log n}{\log p})\}\log p$$

$$= O(n) + \pi\Big(O(n)\Big)O(\log n) = O(n).$$

Here we used the elementary upper estimate $\pi(x)=O(\frac{x}{\log x})$ for the prime number function. This completes the proof.

Since $\sum_{n=0}^{\infty} c_n x^n$ for $c_n = \frac{[\alpha,n][\beta,n]}{[\gamma,n][1,n]}$ is the hypergeometric series, we shall speak of the functions y defined in this section as hypergeometric E-functions. Performing the substitution $x \to \lambda x$ for arbitrary algebraic λ and taking any polynomial in x and finitely many hypergeometric E-functions, with algebraic coefficients, we get again an E-function satisfying a homogeneous linear differential equation whose coefficients are rational functions of x. It would be interesting to find out whether all such E-functions can be constructed in the preceding manner.

We shall apply our theorem to the particular function

$$(89)\qquad K = K_\lambda(x) = \sum_{n=0}^{\infty} \frac{(-1)^n}{n!(\lambda+1)(\lambda+2)\dots(\lambda+n)}\left(\frac{x}{2}\right)^{2n}$$

$$(\lambda \neq -1, -2, \dots).$$

This is a special case of the hypergeometric E-functions introduced in §9: Choose $g=0$, $m=2$, $b_1=1$, $b_2=\lambda+1$ and substitute $\frac{ix}{2}$ for x. The differential equation $W=0$ then becomes

$$K'' + \frac{2\lambda+1}{x}K' + K = 0,$$

so that the two E-functions $y_1 = K$, $y_2 = K'$ are solutions of the system

$$y_1' = y_2, \quad y_2' = -y_1 - \frac{2\lambda+1}{x}y_2.$$

We have to investigate whether these functions satisfy the normality condition in the theorem. This requires a lengthy discussion which, however, is not without interest in itself. The answer will be given in §13, and it will turn out that the normality condition is satisfied for all rational $\lambda \neq \pm\frac{1}{2}, \pm\frac{3}{2}, \dots$.

It is practical to consider the differential equation for $x^\lambda K$ instead of K. This gives the Bessel differential equation

$$(90)\qquad y'' + \frac{1}{x}y' + \left(1 - \frac{\lambda^2}{x^2}\right)y = 0$$

with the particular solution

$$(91)\qquad J_\lambda(x) = \frac{1}{\Gamma(\lambda+1)}\left(\frac{x}{2}\right)^\lambda K_\lambda(x),$$

the Bessel function of index λ. For our next purposes, the parameter λ in (90) may be any complex number, not necessarily rational. Let any solution y of (90) be given, not identically 0. This function is everywhere regular, with possible exception of the points 0 and ∞. It is our first aim to prove that the three functions x, y, y' are not related by an algebraic equation with constant coefficients, except when 2λ is odd.

We start by proving the rather trivial statement that y is not an algebraic function of x. Otherwise there would exist an expansion

$$y = \sum_{k=0}^{\infty} c_k x^{r_k}$$

with decreasing rational exponents $r_0 > r_1 > \dots$ and $c_0 \neq 0$, converging near $x = \infty$. Inserting into (90) we obtain the contradiction $c_0 = 0$.

Now consider more generally an arbitrary homogeneous linear differential equation of the second order

$$w'' + A(x)w' + B(x)w = 0 \tag{92}$$

whose coefficients A and B belong to a given field L of analytic functions of x. We shall assume that L is closed with respect to differentiation, in other words, that L contains the derivatives of all its elements. For instance, L may be the field of rational functions of x. Now suppose that (92) has a particular solution w_0, which is not algebraic over L, but satisfies an algebraic differential equation of the first order, with coefficients in L. We shall prove that then there exists also a solution of (92) whose logarithmic derivative is algebraic over L. Our assumption means that one can find a polynomial P(y,z) in two indeterminates y,z, with coefficients in L and not all 0, such that $P(w_0, w_0') = 0$ identically in x; moreover, P is not independent of z itself.

If Q(y,z) is any polynomial in y,z with coefficients in L, we define

$$(93) \quad Q^*(y,z) = Q_x + zQ_y - (Az+By)Q_z;$$

then $Q^*(y,z)$ again is a polynomial in y,z with coefficients in L and

$$(94) \qquad \frac{d\ Q(w,w')}{dx} = Q^*(w,w')$$

for every solution w of (92). We may assume that P(y,z) is irreducible. Consider P(y,z) and $P^*(y,z)$ as polynomials in z alone and introduce their resultant R(y); this is a polynomial in y with coefficients in L. Because of (94), the differential equation $P(w_0, w_0') = 0$ implies $P^*(w_0, w_0') = 0$, hence $R(w_0) = 0$. But w_0 is not algebraic over L, so that R(y) vanishes identically in y. This proves that P and P^*, as polynomials in z, are not coprime. Since P is irreducible, we obtain

$$(95) \qquad P^*(y,z) = T(y,z)\ P(y,z)$$

where T is a polynomial in y,z with coefficients in L.

Now take in P(y,z) the aggregate H(y,z) of terms of highest total degree in y,z, so that H is a homogeneous polynomial in y,z of positive degree t, with coefficients in L and not all 0. The definition (93) shows that H^* again is a homogeneous polynomial in y,z of degree t, and that $P^*-H^*=(P-H)^*$ is of total degree $< t$. Comparing degrees in (95) we see that T cannot have positive total degree; therefore T is independent of y,z and a function in L; moreover

(96) $$H^*(y,z) = T\,H(y,z).$$

Consider the homogeneous linear differential equation

(97) $$v' = Tv$$

of the first order. Its general solution is $v=cv_0$ where $v_0 \neq 0$ is a particular solution and c an arbitrary constant. Because of (94) and (96), the function $H(w,w')$ is a solution of (97), for <u>every</u> w satisfying (92). Choose two independent solutions w_1, w_2 of (92), then $w = \lambda_1 w_1 + \lambda_2 w_2$ is the general solution of (92), with arbitrary constants λ_1, λ_2. Therefore

$$H(\lambda_1 w_1 + \lambda_2 w_2,\ \lambda_1 w_1' + \lambda_2 w_2') = c(\lambda_1, \lambda_2) v_0,$$

where $c(\lambda_1, \lambda_2)$ is independent of x. But the left-hand side is a homogeneous polynomial of degree t in λ_1, λ_2; hence

$$c(\lambda_1, \lambda_2) = c_0 \lambda_1^{\,t} + c_1 \lambda_1^{\,t-1} \lambda_2 + \dots + c_t \lambda_2^{\,t}$$

with constant $c_0, \dots, c_t$.

Finally determine two constant values λ_1, λ_2, not both 0, such that $c(\lambda_1, \lambda_2) = 0$. Then the corresponding particular solution $w = \lambda_1 w_1 + \lambda_2 w_2$ of (92) satisfies the differential equation

$$H\left(1, \frac{w'}{w}\right) = 0.$$

This means that the logarithmic derivative of w is algebraic over L.

We apply the result of §10 to the Bessel differential equation (90) and take for L the field of rational functions of x. Suppose that $y_0 \neq 0$ is a solution of (90) which satisfies an algebraic differential equation $P(y_0, y'_0)=0$ whose coefficients are polynomials in x and not all 0. It follows from our result that then (90) has another particular solution $y \neq 0$ whose logarithmic derivative $\frac{y'}{y} = u$ is an algebraic function of x. The function y is regular for all $x \neq 0, \infty$; therefore the only possible branch points of u lie at 0 and ∞. We shall prove that no branch of u is ramified at ∞. This implies that also 0 is no branch point, and u has to be a rational function of x.

Let

$$u = \sum_{k=0}^{\infty} c_k x^{r_k} \tag{98}$$

be the power series expansion of any branch of u near $x=\infty$, with rational exponents $r_0 > r_1 > \ldots$ and $c_0 \neq 0$. Because of (90), the function u satisfies the special Riccati differential equation

$$u' + u^2 + \frac{1}{x} u = \frac{\lambda^2}{x^2} - 1;$$

hence

$$\sum_{k=0}^{\infty} (r_k+1) c_k x^{r_k - 1} + \sum_{k,l=0}^{\infty} c_k c_l x^{r_k + r_l} = \frac{\lambda^2}{x^2} - 1. \tag{99}$$

Comparing coefficients we obtain

$$r_0 = 0, \quad c_0 = \pm i.$$

For any $n=1,2,\ldots$ the two terms corresponding to $k=0$, $l=n$ and $k=n$, $l=0$ in the double sum give the contribution

$2c_0c_nx^{r_n}$ to the left-hand side of (99). If $c_n \neq 0$, the exponent r_n has to equal one of the exponents r_k-1 $(k < n)$, r_k+r_l $(k,l < n)$ and -2. We infer by induction that we may restrict the exponents to the sequence of rational integers $r_k=-k$ $(k=0,1,\ldots)$. In particular, $r_1=-1$ and

$$2c_0c_1 + c_0 = 0, \quad c_1 = -\frac{1}{2}.$$

Therefore

$$u = \pm i - \frac{1}{2x} + \ldots \tag{100}$$

is regular and unramified at ∞.

Now we know that u is a rational function of x. Apply (98) and (99) to the power series expansion of u near $x=0$, so that the exponents r_k are consecutive integers in ascending order. It follows that this power series takes the form

$$u = \pm \frac{\lambda}{x} + \ldots . \tag{101}$$

If $x_0 \neq 0, \infty$ is a zero of y, of order $a \geq 1$, then u has at $x=x_0$ a pole of first order with the residue a. Since u is a rational function, y has only finitely many zeros $\neq 0, \infty$, say $x_1,\ldots, x_h$, taken with their multiplicity. The function u is regular at all other points $\neq 0$. It follows from (100) and (101) that

$$u = \pm i \pm \frac{\lambda}{x} + \sum_{k=1}^{h} \frac{1}{x-x_k}$$

is the partial fraction expansion of u and that

$$h \pm \lambda = -\frac{1}{2}.$$

Therefore $2\lambda = \pm(2h+1)$ is an odd number.

We have reached the first aim announced in §10, namely, we have proved that no solution $y \neq 0$ of the Bessel differential equation satisfies an algebraic differential equation of first order whose coefficients are polynomials in x, provided 2λ is not an odd number. It is known that the case $2\lambda = \pm(2h+1)$ really is exceptional. For any given $h=0,1,\ldots$, the functions

$$y_1 = x^{h+\frac{1}{2}} \frac{d^h}{d(x^2)^h} \frac{e^{ix}}{x},$$

$$y_2 = x^{h+\frac{1}{2}} \frac{d^h}{d(x^2)^h} \frac{e^{-ix}}{x}$$

are two independent solutions of (90), with $\lambda = \pm(h+\frac{1}{2})$. Every solution then takes the form

$$y = x^{-h-\frac{1}{2}} (Ae^{ix}+Be^{-ix})$$

where A(x) and B(x) are certain polynomials. Computing $y^2, yy', (y')^2$ and eliminating e^{2ix}, e^{-2ix} one obtains the quadratic differential equation of first order

$$C_1y^2 + C_2yy' + C_3(y')^2 = C_4$$

where C_1, C_2, C_3, C_4 are polynomials in x and not all 0.

From now on we shall exclude the exceptional case.

§12. Algebraic relations involving different Bessel functions

Let two independent solutions y_1, y_2 of the Bessel differential equation be given. Then the function $y_1y_2' - y_2y_1' = \Delta$ satisfies $\Delta' = -\frac{\Delta}{x}$, whence

$$y_1 y_2' - y_2 y_1' = \frac{c}{x} \tag{102}$$

with constant $c \neq 0$. This shows that the five functions y_1, y_1', y_2, y_2' and x are algebraically dependent.

We shall prove that the four functions y_1, y_1', y_2, x are algebraically independent. Because of the result of §11 we have to prove that y_2 is not algebraic over the field M of the rational functions of y_1, y_1', x. Suppose that there exists an irreducible polynomial

$$P(t) = t^n + \ldots$$

with coefficients in M, such that

$$P(y_2) = 0$$

identically in x. We define

$$P^*(t) = \left(\frac{y_1'}{y_1} t + \frac{c}{xy_1}\right) P_t(t) + \frac{dP(t)}{dx} \tag{103}$$

$$= n\frac{y_1'}{y_1} t^n + \ldots ;$$

this again is a polynomial of degree n in t, with coefficients in M. For arbitrary constant λ_1 the function $y_0 = y_2 + \lambda_1 y_1$ is a solution of the linear differential equation of first order

$$y_0' = \frac{y_1'}{y_1} y_0 + \frac{c}{xy_1}$$

for y_0. The definition (103) implies

$$P^*(y_0) = \frac{dP(y_0)}{dx} \tag{104}$$

and, in particular,

$$P^*(y_2) = 0.$$

It follows that $P^*(t)$ is divisible by $P(t)$, whence

$$P^*(t) = n\frac{y_1'}{y_1} P(t)$$

and, because of (104),

$$P(y_0) = by_1^n$$

where b is independent of x. The function $P(y_2+\lambda_1 y_1)$ is a polynomial in λ_1 of degree n; therefore $b=b_0+b_1\lambda_1+\ldots+b_n\lambda_1^n$ with constant $b_0,\ldots,b_n$, and

$$P_t(y_2) = b_1 y_1^{n-1}.$$

This is an algebraic equation for y_2 of degree n-1, with coefficients in M; hence n=1 and y_2 itself lies in M, so that

$$y_2 = \frac{f}{g} \qquad (105)$$

where f and g are polynomials in y_1, y_1' and x. Take in f and g the homogeneous aggregates of highest total degree in y_1, y_1', say f_0 of degree φ and g_0 of degree γ, and introduce the difference $\delta = \varphi - \gamma$ as total degree of the ratio f/g.

Put $f_0/g_0 = v$; then v is homogeneous of degree δ in y_1, y_1', and the difference $f/g-v$ has total degree $< \delta$. Because of the differential equation (90), the same is true for v' and $(f/g)'-v'$. Inserting (105) into (102) we obtain an identity in y_1, y_1', x. On the left-hand side of (102) the terms of highest total degree yield $y_1v'-vy_1'$ of degree $\delta+1$, whereas the right-hand side has total degree 0. Therefore $\delta \geq -1$.

Suppose first that $\delta > -1$; then $y_1v' - vy_1' = 0$, whence $v = c_1y_1$ with constant c_1, and $\delta = 1$. Substituting $y_2 + c_1y_1$ for y_2, we have to replace f/g by $f/g + v$, and the new f/g will have total degree < 1. This leaves us with the case $\delta = -1$ and

$$y_1v' - vy_1' = \frac{c}{x},$$

so that v is a rational function of x, y_1, y_1' and homogeneous of degree -1 in y_1, y_1' which satisfies the Bessel differential equation. Write $v = v(y_1, y_1')$. Using once more the algebraic independence of x, y_1, y_1' we see that also $v(\lambda_1y_1 + \lambda_2y_2, \lambda_1y_1' + \lambda_2y_2')$, for arbitrary constant λ_1 and λ_2, is a solution of (90). Hence

$$(106) \qquad v(\lambda_1y_1 + \lambda_2y_2, \lambda_1y_1' + \lambda_2y_2') = \Lambda_1y_1 + \Lambda_2y_2,$$

where Λ_1, Λ_2 are independent of x. Since

$$vy_2' - y_2v' = \frac{c\Lambda_1}{x}, \qquad y_1v' - vy_1' = \frac{c\Lambda_2}{x},$$

it follows that Λ_1 and Λ_2 are rational functions of λ_1 and λ_2, homogeneous of dimension -1, not both identically 0, and with constant coefficients. Let Λ_0 be the least common denominator of Λ_1 and Λ_2; this is a homogeneous polynomial of degree ≥ 1. Now multiply (106) by Λ_0g_0 and choose for λ_1, λ_2 two numbers, not both 0, such that $\Lambda_0 = 0$. It follows that the particular solution $y = \lambda_1y_1 + \lambda_2y_2$ of (90) satisfies the algebraic differential equation $g_0(y, y') = 0$, which is impossible.

§13. The normality condition for Bessel functions

Now we are going to prove finally that the

normality condition of the theorem is fulfilled in case of the two E-functions $E_1=K_\lambda(x)$, $E_2=K'_\lambda(x)$, where $K_\lambda(x)$ is defined in (89) and λ denotes a rational number $\neq \pm\frac{1}{2}, -1, \pm\frac{3}{2}, -2, \ldots$. We have $E_1'=E_2$, $E_2'=-E_1-\frac{2\lambda+1}{x}E_2$; therefore the q+1 functions $z_{kq}=E_1^k E_2^{q-k}$ $(k=0,\ldots,q)$ satisfy for every given $q=0,1,2,\ldots$ the system of differential equations

$$(107)\qquad z'_{kq} = kz_{k-1,q} - (q-k)\frac{2\lambda+1}{x}z_{kq} - (q-k)z_{k+1,q}$$

$$(k=0,\ldots,q)$$

where $z_{-1,q}$ and $z_{q+1,q}$ mean 0. Our problem is to prove that the $\frac{1}{2}(\nu+1)(\nu+2)$ functions $z_{kq}(k=0,\ldots,q;\ q=0,\ldots,\nu)$ constitute for every $\nu=0,1,2,\ldots$ a normal system.

If z is any solution of

$$(108)\qquad z'' + \frac{2\lambda+1}{x}z' + z = 0$$

then the q+1 functions $z^k(z')^{q-k}$ $(k=0,\ldots,q)$ form a solution of (107). Take $z=\rho_1 z_1+\rho_2 z_2$, where z_1, z_2 are two independent solutions of (108) and ρ_1, ρ_2 arbitrary constants, and define

$$z^k(z')^{q-k} = \sum_{l=0}^{q}\rho_1^l\rho_2^{q-l}\psi_{kl,q},$$

then the q+1 functions $\psi_{kl,q}$ $(k=0,\ldots,q)$ constitute for each $l=0,\ldots,q$ a solution of (107). Since $\psi_{ql,q}=\binom{q}{l}z_1^l z_2^{q-l}$, those q+1 solutions are linearly independent, because otherwise one would obtain a homogeneous algebraic equation of degree q for z_1, z_2, with constant coefficients not all 0, in contradiction to the independence of z_1 and z_2.

The system of $\frac{1}{2}(\nu+1)(\nu+2)$ homogeneous linear differential equations of first order for the z_{kq} $(k=0,\dots,q;\ q=0,\dots,\nu)$ trivially breaks up into $\nu+1$ boxes corresponding to the $\nu+1$ values of q. We obtain the $\nu+1$ solution boxes $Y_q=(\psi_{kl,q})$, where $k,l=0,\dots,q$, and we now have to prove that they are independent in the sense of §4; this means that the sum

$$R = \sum_{0\leq k,l\leq q\leq\nu} P_{kq}\, c_{lq}\, \psi_{kl,q}$$

with arbitrary polynomials $P_{kq}(x)$ and arbitrary constants c_{lq} vanishes identically in x only in the trivial case that all $P_{kq}c_{lq}$ are identically 0.

Introduce $y_1=x^{\lambda} z_1$, $y_2= x^{\lambda} z_2$, so that y_1, y_2 are independent solutions of the Bessel differential equation (90) and

$$z_j' = x^{-\lambda} (y_j'-\frac{\lambda}{x}y_j) \quad (j=1,2), \qquad y_1y_2' - y_2y_1' = \frac{c}{x}$$

with constant $c\neq 0$. Then

$$\rho_1 z_1 + \rho_2 z_2 = x^{-\lambda} (\rho_1 y_1 + \rho_2 y_2),$$

$$\rho_1 z_1' + \rho_2 z_2' = x^{-\lambda} \{\rho_1 y_1' + \rho_2 y_2' - \frac{\lambda}{x}(\rho_1 y_1 + \rho_2 y_2)\}$$

$$= x^{-\lambda} \{(\frac{y_1'}{y_1} - \frac{\lambda}{x})(\rho_1 y_1 + \rho_2 y_2)+\frac{c\rho_2}{xy_1}\}$$

and

$$(109) \qquad \sum_{l=0}^{q} \rho_1^{\,l} \rho_2^{\,q-l} \psi_{kl,q}$$

$$= x^{-\lambda q}(\rho_1 y_1 + \rho_2 y_2)^k \{(\frac{y_1'}{y_1} - \frac{\lambda}{x})(\rho_1 y_1 + \rho_2 y_2)+\frac{c\rho_2}{xy_1}\}^{q-k}$$

This shows that $\psi_{kl,q}$ is a polynomial in y_1, y_1^{-1}, y_1', y_2 whose coefficients are algebraic functions of x, since λ is a rational number.

Now suppose that R=0, identically in x. The result of §12 implies that then R vanishes identically in x, y_1, y_1', y_2. If all products $P_{kq}c_{lq}=0$ for $k,l=0,\ldots,q$ and $q=\mu+1,\ldots,\nu$, we consider in R the terms of total degree μ in y_1, y_1', y_2. Because of (109), we obtain from $\psi_{kl,\mu}$ the contribution $\binom{\mu}{l}x^{-\lambda q}$ $\left(\frac{y_1'}{y_1}-\frac{\lambda}{x}\right)^{\mu-k} y_1^l\, y_2^{\mu-l}$. Therefore

$$\sum_{0\leq k,l\leq\mu} P_{k\mu}\, c_{l\mu}\, \binom{\mu}{l}\left(\frac{y_1'}{y_1}-\frac{\lambda}{x}\right)^{\mu-k}\left(\frac{y_2}{y_1}\right)^{\mu-l} = 0$$

identically in x, y_1, y_1', y_2, and this implies $P_{k\mu}\, c_{l\mu} = 0$ $(k,l=0,\ldots,\mu)$. It follows that all P_{kq} c_{lq} are 0, and the proof is complete.

The only singularity of the coefficients in the differential equation (108), for finite x, is x=0. Let α be any algebraic number $\neq 0$, and λ a rational number $\neq \pm\frac{1}{2}, -1, \pm\frac{3}{2}, -2, \ldots$. It follows from the theorem that the values $K_\lambda(\alpha)$ and $K_\lambda'(\alpha)$ are not related by any algebraic equation with algebraic coefficients. In particular, $K_\lambda(\alpha)$ itself is transcendental. This implies that all zeros of $K_\lambda(x)$ are transcendental, and, because of (91), the same holds for the zeros $\neq 0$ of the Bessel function $J_\lambda(x)$.

Putting

$$w = f_\lambda(x) = \sum_{n=0}^{\infty} \frac{x^n}{n!(\lambda+1)\ldots(\lambda+n)}$$

we have

$$K_\lambda(x) = f_\lambda\left(-\frac{x^2}{4}\right)$$

and the differential equation

$$xw'' + (\lambda + 1)w' = w,$$

which leads to the continued fraction

$$\frac{w}{w'} = \lambda + 1 + \cfrac{x}{\lambda + 2 + \cfrac{x}{\lambda + 3 + \ddots}}$$

Since

$$\frac{w'}{w} = \frac{-1}{\sqrt{-x}} \frac{K'(2\sqrt{-x})}{K(2\sqrt{-x})},$$

it follows that the value of this continued fraction is transcendental for all algebraic $x \neq 0$ and all rational λ. In this statement the values $\lambda = \pm\frac{1}{2}, \pm\frac{3}{2}, \ldots$ are included, because the remark at the end of §11 shows that in this case the transcendency follows from that of $e^{\sqrt{x}}$.

In particular, the number

$$1 + \cfrac{1}{2 + \cfrac{1}{3 + \ddots}}$$

is transcendental.

§14. Additional remarks

The preceding example of the Bessel differential equation makes clear that the application of the theorem to a given system of linear differential equations of first order requires a detailed study of algebraical and analytical properties of the solutions. Of course, it may happen that the normality condition is not satisfied. The simplest example is provided by the special case $\lambda = -\frac{1}{2}$ in (89); then $K=\cos x$, $K' = -\sin x$, $K'' = -K$, and the solution box

$$Y = \begin{pmatrix} \cos x & \sin x \\ -\sin x & \cos x \end{pmatrix}$$

is not independent in the sense of §4, since $\cos x + \sin x \cdot i + i(-\sin x) + i \cos x \cdot i = 0$. The inner reason is that the substitution $z_1=y_1-iy_2$, $z_2=y_1+iy_2$ transforms the system $y_1'=y_2$, $y_2'=-y_1$ into $z_1'=iz_1$, $z_2'=iz_2$, and the transformed Y breaks up into the two boxes e^{ix}, e^{-ix}. This makes it probable that the independence of the solution boxes Y_t in §4 can be expressed as a property of the ring P of linear transformations

$$y_k \longrightarrow A_{k1} y_1 + \dots + A_{km} y_m \qquad (k=1,\dots,m),$$

whose coefficients A_{kl} are rational functions of x, and which map the linear set of solutions of the differential equations (49) into itself. It is not difficult to find this connection if one uses the results of A. Loewy on homogeneous linear differential equations. However, we did not need **P** for the application to Bessel functions, so we only mention it here.

Our result concerning $K(\alpha)$ can be generalized in the following way. We consider the $m=2r$ functions $K(\alpha_l x)$, $K'(\alpha_l x)$ $(l=1,\dots,r)$, where $\alpha_1,\dots,\alpha_r$ are given algebraic numbers whose squares all are different from each other and from 0. It is possible to prove that the condition of the theorem is satisfied, if 2λ is not odd; therefore the numbers $K(\alpha_1)$, $K'(\alpha_1),\dots,K(\alpha_r)$, $K'(\alpha_r)$ are not related by an algebraic equation with algebraic coefficients. This again is a special case of the following statement: Let $\lambda_1, \ldots, \lambda_s$ be rational numbers, different from $-1, \pm\frac{1}{2}, -2, \pm\frac{3}{2}, \ldots$, and let none of the numbers $\lambda_k \pm \lambda_l$ $(1 \leq k < l \leq s)$ be integral; then the 2rs numbers $K_\lambda(\alpha)$, $K'_\lambda(\alpha)$ $(\lambda = \lambda_1, \ldots, \lambda_s;$ $\alpha = \alpha_1, \ldots, \alpha_r)$ are algebraically independent. The proof has not been carried through in detail and we leave it as an interesting exercise.

Another exercise is provided by the hypergeometric E-function

$$y = 1 + \frac{\kappa}{\lambda} \cdot \frac{x}{1!} + \frac{\kappa(\kappa+1)}{\lambda(\lambda+1)} \cdot \frac{x^2}{2!} + \cdots$$

$$= \int_0^1 t^{\kappa-1}(1-t)^{\lambda-\kappa-1}e^{tx}dt \Big/ \int_0^1 t^{\kappa-1}(1-t)^{\lambda-\kappa-1}dt$$

with rational κ, $\lambda \neq 0, -1, -2, \ldots$, which is a solution of

$$xy'' + (\lambda - x)y' = \kappa y.$$

It seems more difficult to carry over the investigations of Sections 10, 11, 12, 13 to linear differential equations of order greater than two, but it would be worth while to try.

CHAPTER III

THE TRANSCENDENCY OF a^b

FOR IRRATIONAL ALGEBRAIC b AND ALGEBRAIC $a \neq 0, 1$

Let ρ be any complex number $\neq 0$ such that $e^{\rho} = a$ is algebraic, and let β be any irrational number such that also $e^{\rho\beta} = a^{\beta} = c$ is algebraic. Because of the addition theorem $f(x+y) = f(x)f(y)$ for the function $f(x) = e^{\rho x} = a^x$ it follows that $f(x)$ takes the algebraic values $a^p c^q$ for all $x = p + \beta q$ $(p, q = 0, \pm 1, \pm 2, \ldots)$. These x are different from each other, because of the irrationality of β.

In 1929 Gelfond made the important discovery that β cannot be a non-real quadratic irrationality; this means that the number $a^b = e^{b \log a}$ is transcendental for every imaginary quadratic irrationality b and every algebraic $a \neq 0$, except for $\log a = 0$. In his proof Gelfond applied the interpolation formula of §14, Chapter I, to the special case $f(z) = a^z = e^{z \log a}$ and the sequence $z_1, z_2, \ldots$ consisting of the points $p + bq (p, q = 0, \pm 1, \pm 2, \ldots)$ in a certain arrangement. The coefficients $a_0, a_1, \ldots$ in the expansion

$$f(z) = a_0 F_0(z) + a_1 F_1(z) + \ldots,$$

$$F_n(z) = \prod_{l=1}^{n} (z - z_l)$$

are given by

$$a_{n-1} = \frac{1}{2\pi i} \int_C \frac{f(z)}{F_n(z)}\, dz = \sum_{k=1}^{n} \frac{f(z_k)}{F_n'(z_k)},$$

$$F_n'(z_k) = \prod_{\substack{l=1 \\ l \neq k}}^{n} (z_k - z_l),$$

and this shows that they are numbers in the field K generated by a, b, a^b. On one hand it follows from the integral formula for a_{n-1} that the coefficients tend rapidly to 0 as $n \to \infty$; on the other hand, if K were an algebraic number field, one can obtain an estimate of $\overline{|a_n|}$ and the denominator of a_n. In case of an imaginary quadratic b these estimates imply that $a_n = 0$ for all sufficiently large n. This is contradictory, since f(z) is not a polynomial.

Gelfond's proof was carried over to the case of a real quadratic irrationality b by Kusmin in 1930. However, this method fails if b is an algebraic irrationality of degree $h > 2$; it only gives the weaker result that at least one of the h-1 numbers $a^b, a^{b^2}, \ldots, a^{b^{h-1}}$ is transcendental.

In 1934 Gelfond and Schneider, independently of each other, solved the general problem of the transcendency of a^b for arbitrary irrational algebraic b. Both proofs make use of the arithmetical lemmas in §2, Chapter II, in order to construct suitable approximation forms. Since Schneider's proof is more closely related to the other ideas of Chapter II, we present it first. Gelfond's proof will be given in a simplified form which makes it somewhat shorter than Schneider's proof.

§1. Schneider's proof

Suppose that b is irrational and that the three

numbers b, $a \neq 0$, $c = a^b = e^{b \log a}$ ($\log a \neq 0$) lie in an algebraic number field K of degree h. If a is a root of unity, then c is not a root of unity; in this case, we replace a, b, c by c, b^{-1}, a, so that the new a is not a root of unity.

Define

$$m = 4h + 3, \qquad n = \frac{q^2}{m},$$

where q^2 is a positive integral rational square divisible by m. We now apply lemma 2, §2, Chapter II, to the determination of m polynomials $P_1(x), \ldots, P_m(x)$ of degree $\leq 2n-1$ with integral coefficients in K, not all 0, such that the entire function

$$R(x) = P_1(x)a^{(m-1)x} + P_2(x)a^{(m-2)x} + \ldots + P_m(x)$$

vanishes at the $q^2 = mn$ different points $x = \lambda + \mu b$ ($\lambda, \mu = 1, 2, \ldots, q$). This condition yields mn homogeneous linear equations for the 2mn indeterminate coefficients in $P_1, \ldots, P_m$. The numerical coefficients of these equations lie in K and the absolute values of all their conjugates are $(c_1 q)^{2n-1} O(c_2^{\,q}) = O(c_3^{\,n} n^n)$, where c_1, $c_2, \ldots$ denote positive rational integers independent of n; moreover, there exists a common denominator $c_4^{\,2n-1+2(m-1)q} = O(c_5^{\,n})$. In virtue of the lemma, we obtain a non-trivial solution such that all coefficients in $P_1, \ldots, P_m$, and their conjugates, are $O(c_6^{\,n} n^n)$.

Putting

$$P_{k1}(x) = a^{(m-1)k} P_1(x+k) \qquad (k = 1, \ldots, m), \tag{110}$$

we have

(111) $$R(x+k) = P_{k1}a^{(m-1)x} + \dots + P_{km}.$$

Suppose that $P_{\nu_1}(x),\dots,P_{\nu_g}(x)$ $(0 < \nu_1 < \nu_2 < \dots < \nu_g \leq m)$ are not identically 0, whereas all other $P_l(x)=0$. If

$$P_{\nu_l} = \alpha_l x^{r_l} + \dots, \qquad \alpha_l \neq 0 \qquad (l=1,\dots,g),$$

so that $r_l \geq 0$ denotes the exact degree of P_{ν_l}, then the g-rowed determinant

$$\Delta(x) = | P_{k\nu_l}(x) |_{k,l=1,\dots,g}$$

$$= \alpha_1 \dots \alpha_g v x^{r_1+\dots+r_g} + \dots,$$

with

$$v = | a^{(m-\nu_l)k} |_{k,l=1,\dots,g}$$

$$= \prod_{k>l} (a^{m-\nu_k} - a^{m-\nu_l}) \neq 0,$$

does not vanish identically, since a is not a root of unity. The degree of $\Delta(x)$ is $r_1+\dots+r_g \leq m(2n-1) < 2mn$. Therefore we can find at least one of the $2q^2=2mn$ numbers

$$x = \lambda + \mu b \qquad (\lambda = 1,\dots,2q;\ \mu = 1,\dots,q),$$

say $x=\xi$, such that

$$\Delta(\xi) \neq 0.$$

It follows from (110) and (111) that at least one of the m numbers $R(\xi+k)$ $(k=1,\ldots,m)$ is different from 0. Suppose

$$R(\xi+k) = \gamma \neq 0,$$

then $c_4^{2n-1+(m-1)(3q+m)}\gamma$ is an integer in K and $\neq 0$; whence

$$(112) \qquad |\mathfrak{N}(\gamma)| > c_7^{-n}.$$

On the other hand, using our previous estimate of the coefficients of $P_1,\ldots,P_m$, we obtain

$$(113) \qquad \overline{|\gamma|} = 2mn(c_8 q)^{2n-1} c_9^q \; O(c_6^n n^n) = O(c_{10}^n n^{2n}).$$

It remains to find a sufficiently good upper estimate of $|\gamma|$ itself. Apply Cauchy's theorem to the <u>entire</u> function

$$S(x) = R(x) \prod_{\lambda,\mu=1}^{q} \frac{\xi+k-\lambda-\mu b}{x-\lambda-\mu b}$$

which takes the value γ at $x=\xi+k$; then

$$\gamma = \frac{1}{2\pi i}\int_C \frac{S(x)}{x-\xi-k}\,dx,$$

where C is a simple closed curve, in positive direction,

containing $\xi+k$ in its interior. Taking for C the circle

$$|x| = n + m + q(2+|b|) = n + O(\sqrt{n})$$

we have

$$R(x) = 2mn(c_{11}n)^{2n-1}c_{12}{}^n O(c_6{}^n n^n) = O(c_{13}{}^n n^{3n}),$$

$$|x-\xi-k| \geq n > \frac{|x|}{c_{14}}, \qquad |x-\lambda-\mu b| > n \ (\lambda, \mu=1,\dots,q)$$

$$\prod_{\lambda,\mu=1}^{q} \frac{\xi+k-\lambda-\mu b}{x-\lambda-\mu b} = n^{-q^2} O\left((c_{15}q)^{q^2}\right) = O(c_{16}{}^n n^{-\frac{mn}{2}}),$$

hence

$$(114) \qquad \gamma = c_{13}{}^n n^{3n} O(c_{16}{}^n n^{-\frac{mn}{2}}) = O\left(c_{17}{}^n n^{(3-\frac{m}{2})n}\right).$$

In view of (113) and (114),

$$(115) \qquad \mathfrak{N}(\gamma) = O\left(c_{18}{}^n n^{(3-\frac{m}{2})n+2(h-1)n}\right).$$

Here the exponent of n equals

$$(3-\frac{m}{2})\,n + 2(h-1)n = -\frac{n}{2},$$

so that (112) and (115) contradict each other as $n \to \infty$.

§2. Gelfond's proof

We use a, b, c, h, K in the same meaning as in §1: The three numbers a, b and $c=a^b=e^{b \log a}$ lie in an algebraic number field of degree h; the number b is irrational, a and log a are $\neq 0$. However, we need not assume that a is not a root of unity.

The approximation form is constructed in a

different way. Put

$$m = 2h + 2, \qquad n = \frac{q^2}{2m},$$

where $q^2=t$ is a positive integral rational square divisible by 2m, and

$$\rho_1, \rho_2, \ldots, \rho_t = (\lambda + \mu b) \log a \qquad (\lambda, \mu = 1, \ldots, q).$$

Introduce the entire function

$$R(x) = \eta_1 e^{\rho_1 x} + \ldots + \eta_t e^{\rho_t x}$$

with indeterminate coefficients $\eta_1, \ldots, \eta_t$, and consider the mn homogeneous linear equations

$$(\log a)^{-k} R^{(k)}(l) = 0 \qquad (k=0, \ldots, n-1;\ l=1, \ldots, m)$$

for the $2mn=t$ unknown quantities $\eta_1, \ldots, \eta_t$. Their numerical coefficients lie in K, they have a number

$$c_1^{\,n-1+2mq} = O(c_2^{\,n})$$

as common denominator, and all their conjugates are $O\left((c_3 q)^{n-1} c_4^{\,q}\right) = O(c_5^n\, n^{n/2})$. It follows from lemma 2 that the equations have a non-trivial solution in integers $\eta_1, \ldots, \eta_t$ of K such that

$$\overline{|\eta_k|} = O(c_6^n\, n^{\frac{n}{2}}) \qquad (k=1, \ldots, t).$$

Since the t numbers ρ_k all are different, the function R(x) does not vanish identically. Choose p such that $R^{(k)}(x)=0$ for $k=0,1,\ldots,p-1$ and $x=1,\ldots,m$, but $R^{(p)}(l) \neq 0$, for some $l=1,\ldots,m$. Clearly, $p \geq n$. Consider the number

$$(116) \qquad (\log a)^{-p} R^{(p)}(l) = \gamma \neq 0;$$

it lies in K, and $c_1^{p+2mq}\gamma$ is integral, so that

$$(117) \qquad |\mathfrak{N}(\gamma)| > c_7^{-p}$$

Moreover,

$$(118) \qquad \overline{|\gamma|} = t(c_3 q)^p c_4^{\,q}\, O(c_6^n\, n^{\frac{n}{2}}) = O(c_8^p\, p^p).$$

To find again a suitable upper estimate of $|\gamma|$ itself, we introduce the <u>entire</u> function

$$S(x) = p!\ \frac{R(x)}{(x-l)^p} \prod_{\substack{k=1 \\ k \neq l}}^{m} \left(\frac{l-k}{x-k}\right)^p;$$

then

$$\gamma = (\log a)^p S(l)$$

and

$$S(l) = \frac{1}{2\pi i} \int_C \frac{S(x)}{x-l}\, dx,$$

where we take for C the circle

$$|x| = m(1+\frac{p}{q}).$$

We obtain

$$R(x) = t\, c_9^{p+q}\, O(c_6^n\, r^{\frac{n}{2}}) = O(c_{10}^p\, p^{\frac{p}{2}}),$$

$$|x-1| \geq |x| \frac{p}{p+q}, \qquad |x-k| \geq m \frac{p}{q} \qquad (k=1,\ldots,m)$$

$$(x-1)^{-p} \prod_{\substack{k=1\\k\neq 1}}^{m} \left(\frac{1-k}{x-k}\right)^p = O\left(c_{11}^p \left(\frac{q}{p}\right)^{mp}\right)$$

$$S(x) = p!\, c_{11}^p \left(\frac{q}{p}\right)^{mp} O(c_{10}^p\, p^{\frac{p}{2}}) = O(c_{12}^p\, p^{\frac{3-m}{2}p});$$

whence

$$\gamma = O\left(c_{13}^p\, p^{\frac{3-m}{2}p}\right)$$

and, by (118),

$$\mathfrak{N}(\gamma) = O(c_{14}^p\, p^{(h-1)p + \frac{3-m}{2}p}). \tag{119}$$

The exponent of p equals

$$(h-1)p + \frac{3-m}{2}p = -\frac{p}{2},$$

so that (117) and (119) contradict each other for sufficiently large n, because of $p \geq n$.

§3. Additional remarks

The main difference of the two proofs lies in the method by which the algebraic number $\gamma \neq 0$ is obtained. Schneider applies the functional equation $a^{x+y} = a^x a^y$ in order to construct g approximation forms with non-vanishing determinant, and the inequality $\gamma \neq 0$ follows from an algebraical reason, in analogy to our procedure in §4 and §5, Chapter II. Gelfond simply uses the fact that the Taylor series of an analytic function, which is not a polynomial, must contain infinitely many coefficients $\neq 0$. Since this method is indirect, it does not give immediately an explicit finite upper bound for p in (116); though this can be found by a more detailed investigation. Gelfond's version of the proof is appropriate for the solution of the analogous problems concerning elliptic functions which we shall study in Chapter IV.

Our result on the transcendency of a^b can also be stated this way: If a and c are algebraic numbers, $ac \neq 0$, $\log a \neq 0$, then the ratio $\log c / \log a$ is either rational or transcendental. In other words: The logarithm of any algebraic number relative to any algebraic base is either rational or transcendental. However, it is not known whether log c and log a are algebraically independent in the case of an irrational quotient $\log c / \log a$; it is not even known whether there cannot exist an inhomogeneous linear relation $\alpha \log a + \gamma \log c = 1$ with quadratic irrational α and γ. Another example showing the narrow limits of our knowledge on transcendental numbers is the following one: Since e is transcendental, not both numbers $e+\pi$ and $e\pi$ can be algebraic; but we do not even know whether $e+\pi$ or $e\pi$ are irrational.

Since $e^{\pi} = i^{-2i}$, the transcendency of e^{π} is contained in Gelfond's result of 1929 on the transcendency of a^b for algebraic $a \neq 0$, $\log a \neq 0$, and imaginary quadratic b. Before this discovery the problem of proving, for instance, the irrationality of $2^{\sqrt{2}}$ had been considered as extremely difficult, so that Hilbert liked to mention it as a problem whose solution lay still further in the future than the proof of Riemann's hypothesis or Fermat's conjecture. This shows that one cannot guess the real difficulties of a problem before having solved it.

CHAPTER IV

ELLIPTIC FUNCTIONS

The last chapter will be devoted to some of the deep results concerning elliptic functions obtained by Schneider in 1937.

§1. Abelian differentials

Let $\varphi(\xi, \eta)=0$ be the equation of an irreducible algebraic curve C of genus p. We consider ξ as the independent variable and introduce the Riemann surface $\mathfrak{R}$ corresponding to the algebraic function η. The field of all single-valued meromorphic functions on $\mathfrak{R}$ is identical with the field Ω of all rational functions $\psi(\xi, \eta)$ of ξ and η on C which are not identically ∞. The expression

$$dw = \psi(\xi, \eta) d\xi \tag{120}$$

is called an <u>abelian differential</u> on $\mathfrak{R}$.

At every point $\mathfrak{p}$ on $\mathfrak{R}$ we have a local uniformizing parameter t which maps a neighborhood of this point onto a simple neighborhood of 0 in the t-plane. Inserting for ξ and η their power series in t, we obtain from (120) an expansion

$$\frac{dw}{dt} = \sum_{n=-\infty}^{\infty} c_n t^n$$

where only finitely many c_n with negative subscript n are $\neq 0$; of course, the coefficients c_n depend upon $\mathfrak{p}$. The abelian differential dw is called of the first kind, if $c_n=0$ for all $n < 0$ and all $\mathfrak{p}$; it is of the second kind, if $c_{-1}=0$ everywhere on $\mathfrak{R}$; in the remaining case, when there is no condition, one speaks of the third kind. Introducing the abelian integrals w one can characterize the three cases in the following way: The abelian integral w is of the first kind, if it is everywhere finite; it is of the second kind, if it has no other singularities on $\mathfrak{R}$ than poles; it is of the third kind, if it may have logarithmic branch points.

If dw_1 and dw_2 are abelian differentials on $\mathfrak{R}$ of the same kind, then also $\lambda_1 dw_1 + \lambda_2 dw_2$ is, for any constants λ_1, λ_2. We shall restrict ourselves to differentials of the first and second kind. The structure of the corresponding additive groups is described by the following classical results: There exist p, and not more than p, linearly independent differentials $du_1, \ldots, du_p$ of the first kind; there exist p differentials of the second kind $dv_1, \ldots, dv_p$ such that any differential of the second kind on $\mathfrak{R}$ can be uniquely expressed in the form

$$(121) \quad dw = (\lambda_1 du_1 + \ldots + \lambda_p du_p) + (\mu_1 dv_1 + \ldots + \mu_p dv_p) + d\chi ,$$

where $\lambda_1, \ldots, \lambda_p$ and $\mu_1, \ldots, \mu_p$ are constants, and $\chi = \chi(\xi, \eta)$ lies in Ω.

In the preceding definitions and statements it is not necessary to assume that the underlying field of constants in Ω comprises all complex numbers, but is is important that this field is algebraically closed. From now on we shall restrict the coefficients of φ and ψ

to algebraic numbers and speak of Ω in this new meaning; then also $\lambda_1, \ldots, \lambda_p$ and $\mu_1, \ldots, \mu_p$ in (121) are algebraic numbers.

§2. Elliptic integrals

In the elliptic case p=1 the curve $\varphi(\xi, \eta)=0$ can be mapped by an appropriate birational transformation

$$x = \psi_1(\xi, \eta), \qquad y = \psi_2(\xi, \eta) \qquad (\psi_1, \psi_2 \text{ in } \Omega)$$

onto the normal cubic curve

$$y^2 = 4x^3 - g_2 x - g_3$$

with algebraic constants g_2, g_3 and $g_2^3 - 27g_3^2 = \Delta \neq 0$. Then

$$dz = -\frac{dx}{y}, \qquad d\zeta = \frac{xdx}{y}$$

are differentials of the first and second kind, and any elliptic integral w of the second kind satisfies, in virtue of (121), the decomposition formula

$$dw = \lambda\, dz + \mu\, d\zeta + d\chi, \tag{122}$$

where λ, μ are algebraic constants and $\chi(x, y)$ is a rational function of x and y with algebraic coefficients.

The Weierstrassian functions $\wp(z)$ and $\zeta(z)$ are defined by

$$z = \int_x^{\infty} \frac{dx}{y}, \qquad x = \wp(z), \qquad y = \wp'(z) = -\frac{dx}{dz},$$

$$\zeta'(z) = \frac{d\zeta}{dz} = \frac{d\zeta}{dx} \cdot \frac{dx}{dz} = -x = -\wp(z), \qquad \zeta(-z) = -\zeta(z).$$

Introduce the odd function

$$q(z) = \lambda z + \mu \zeta (z),$$

then the decomposition (122) takes the form

$$(123) \qquad dw = dq + d\chi .$$

We shall have to make use of the addition theorem for the elliptic integral w of the second kind; because of (123), it is an immediate consequence of the addition theorem for q, namely

$$(124) \qquad q(z+z_0) - q(z) - q(z_0) = \frac{\mu}{2} \frac{\wp'(z) - \wp'(z_0)}{\wp(z) - \wp(z_0)},$$

where z and z_0 are independent variables in the complex plane. The addition theorem for $\wp(z)$ follows from (124) by differentiation with respect to z, in view of the differential equation

$$\wp'(z) = \left(4 \wp^3(z) - g_2 \wp(z) - g_3\right)^{\frac{1}{2}}.$$

Besides (124) we need some other well known properties of the elliptic functions: There exist two basic periods ω_1, ω_2 of $\wp(z)$ such that any period ω can be uniquely expressed as $\omega = g_1 \omega_1 + g_2 \omega_2$ with rational integral g_1, g_2; the ratio $\omega_2/\omega_1 = \tau$ is not real; the function $\zeta(z)$ has poles of first order at all period points $z = \omega$ and is regular at all other finite points in the z-plane; finally,

$$(125) \qquad q(z+\omega) - q(z) = \eta ,$$

where η depends upon the period ω and not upon z.

Our main object is the proof of the following

theorem by Schneider: Let $\mathfrak{p}_1 = \mathfrak{p}(\xi_1, \eta_1)$ and $\mathfrak{p}_2 = \mathfrak{p}(\xi_2, \eta_2)$ be two points on the Riemann surface $\mathfrak{R}$ of the curve $\varphi(\xi, \eta) = 0$ of genus 1, whose coordinates ξ_1, η_1 and ξ_2, η_2 are algebraic numbers, and let $w(\mathfrak{p})$ be an indefinite elliptic integral of the second kind which is regular at $\mathfrak{p}_1, \mathfrak{p}_2$ and does not reduce to a rational function of ξ and η; then the value of the definite elliptic integral

$$(126) \qquad w(\mathfrak{p}_2) - w(\mathfrak{p}_1) = \int_{\mathfrak{p}_1}^{\mathfrak{p}_2} dw$$

is transcendental, except when $\mathfrak{p}_1, \mathfrak{p}_2$ coincide and the path of integration on $\mathfrak{R}$ is homologous to 0.

It should be mentioned, since $w(\mathfrak{p})$ is not single-valued on $\mathfrak{R}$, that the theorem holds for any path of integration L on $\mathfrak{R}$ with the end points $\mathfrak{p}_1$ and $\mathfrak{p}_2$. For the proof of the theorem we may exclude the trivial case that L is closed on $\mathfrak{R}$ and homologous to 0; then L is the image of a path in the z-plane with different end points z_1 and z_2. The difference

$$z_2 - z_1 = z_0 \neq 0$$

is a period ω only in case L is closed on $\mathfrak{R}$. In this case we may suppose, for the proof of the theorem, that L is not homologous to twice a closed curve on $\mathfrak{R}$; then $\frac{\omega}{2}$ is not a period and, because of (123) and (125),

$$(127) \qquad 2q(\tfrac{\omega}{2}) = \eta = \int_L dq = \int_L dw, \qquad \wp'(\tfrac{\omega}{2}) = 0.$$

If L is not closed, then z_0 is not a period and we may apply (124), with $z=z_1$ or z tending to z_1, and (123). By the assumptions in the theorem concerning $\mathfrak{p}_1$ and $\mathfrak{p}_2$, it follows that the difference of $q(z_0)$ and the elliptic integral (126) is an algebraic number. This result, together with (127), shows that it suffices to prove the following statement:

Let λ, μ, g_2, g_3, $\wp(z_0)$ be algebraic numbers and λ, μ not both 0; then $q(z_0) = \lambda z_0 + \mu\zeta(z_0)$ is transcendental.

§3. The approximation form

Suppose that λ, μ, g_2, g_3, $\wp(z_0)$, $\wp'(z_0)$ and $q(z_0)$ lie in an algebraic number field K of degree h and put

$$m = 16h + 1. \tag{128}$$

Consider the multiples $z_0, 2z_0, 3z_0, \ldots$ of z_0; since z_0 is not a period, we may choose m of these multiples, say $z_1, \ldots, z_m$, which are no periods. It follows from the addition theorem of $q(z)$ and $\wp(z)$ that all 3m numbers $\wp(z_k)$, $\wp'(z_k)$, $q(z_k)$ $(k=1,\ldots,m)$ lie in K.

If a and b are any constants $\neq 0$ in K, we may perform the substitutions $a^2x \to x$, $a^3y \to y$, $a^4g_2 \to g_2$, $a^6g_3 \to g_3$, $a^{-1}z \to z$, $ab\lambda \to \lambda$, $a^{-1}b\mu \to \mu$, $a\zeta \to \zeta$, $bq \to q$, which simply express the group property of abelian differentials. A suitable choice of a and b then allows us to assume that the 3m+4 numbers λ, μ, $\frac{1}{2}g_2$, g_3, $\wp(z_k)$, $\wp'(z_k)$, $q(z_k)$ $(k=1,\ldots,m)$ all are integers in K. It follows from the formulas $q' = \lambda - \mu\wp$, $\wp'' = 6\wp^2 - \frac{1}{2}g_2$ that also all derivatives of $\wp$ and q are integers in K, for $z = z_1, \ldots, z_m$, and the same then holds obviously for all derivatives of the power products

$$f(z) = f_{\alpha\beta}(z) = \wp^{\alpha}(z)q^{\beta}(z) \qquad (\alpha, \beta = 0, 1, \ldots).$$

To obtain an upper estimate of the absolute values of these derivatives we use Cauchy's formula

$$f^{(l)}(z_k) = \frac{l!}{2\pi i}\int_C \frac{f(z)}{(z-z_k)^{l+1}}\,dz \qquad (l=0,1,\ldots),$$

taking for C a circle with center z_k and sufficiently small radius, so that all period points stay outside. On C we have $f(z)=O(c_1^{\alpha+\beta})$ where c_1 is independent of α, β; hence

$$f^{(l)}_{\alpha\beta}(z_k) = l!\ O(c_2^{\alpha+\beta+l}) \tag{129}$$

for $k=1,\ldots,m$ and all α, β, $l=0,1,\ldots$.

A corresponding estimate for the conjugates of $f^{(l)}(z_k)$ can be obtained by the following trick. Let $\overline{K}$ be any of the h conjugates of K. Since $\Delta = g_2^3-27g_3^2 \neq 0$, also $\overline{\Delta} \neq 0$; therefore we may introduce the $\wp$-function with the invariants $\overline{g_2}$, $\overline{g_3}$, and we denote it by $\overline{\wp}(z)$. For fixed k we define a complex number $\overline{z_k}$ by the condition that $\overline{\wp}(\overline{z_k})=\overline{\wp(z_k)}$ and $\overline{\wp}'(\overline{z_k})=\overline{\wp'(z_k)}$; this determines $\overline{z_k}$ only up to an arbitrary additive period of the function $\overline{\wp}(z)$, and we choose a fixed $\overline{z_k}$. Finally we define the function $\overline{q}(z)$ uniquely by the condition that $\overline{q}'(z)=\overline{\lambda}-\overline{\mu}\,\overline{\wp}(z)$ and $\overline{q}(\overline{z_k})=\overline{q(z_k)}$. Then $\overline{f}_{\alpha\beta}(z)=\overline{\wp}^{\alpha}\,\overline{q}^{\beta}$ has all the properties of $f_{\alpha\beta}(z)$ needed for the proof of (129), and $\overline{f}^{(l)}_{\alpha\beta}(\overline{z_k})=\overline{f^{(l)}_{\alpha\beta}(z_k)}$. Therefore

$$\left|\overline{f^{(l)}_{\alpha\beta}(z_k)}\right| = l!\ O(c_3^{\alpha+\beta+l}) \tag{130}$$

for $k=1,\ldots,m$ and all α, β, $l=0,1,\ldots$.

Let r^2 be a positive rational integral square divisible by 2m and put

$$n = \frac{r^2}{2m}.$$

Consider the polynomial $R(\wp, q)$ in $\wp$ and q, of degree $< r$ in each variable, with r^2 indeterminate coefficients. Under the condition that $R(\wp, q)$, as a function of z, vanishes at all m points $z = z_1, \ldots, z_m$ of order n at least, we obtain mn homogeneous linear equations with integral numerical coefficients in K whose conjugates, because of the estimate (130) for $\alpha < r$, $\beta < r$, $l < n$, all are $O(c_4^n n^n)$. Applying lemma 2 we see that we can satisfy our condition by a polynomial $R(\wp, q)$ with integral coefficients in K, not all 0, whose conjugates are $O(c_5^n n^n)$.

§4. Conclusion of the proof

We shall first prove that the function q(z) is not an algebraic function of $\wp(z)$. Otherwise there would exist an equation of smallest degree $\rho \geq 1$,

$$q^\rho + A_1 q^{\rho - 1} + \ldots + A_\rho = 0, \tag{131}$$

where $A_1, \ldots, A_\rho$ are rational functions of $\wp(z)$ and $\wp'(z)$. Since $q'(z) = \lambda - \mu \wp(z)$, differentiation of (131) with respect to z gives an equation of degree $\leq \rho - 1$ in q and therefore an identity in q. Let j be an indeterminate, independent of z. It follows that the expression

$$(q+j)^\rho + A_1 (q+j)^{\rho - 1} + \ldots + A_\rho, \quad q = q(z),$$

does not depend upon z. Derivation with respect to j gives an equation of degree $\rho - 1$ for q(z); hence $\rho = 1$, and q(z) is an elliptic function. But q(z) has at most one pole in the period parallelogram, namely the pole of first order at z=0 in case $\mu \neq 0$. Therefore q(z) is constant, and this is a contradiction, λ and μ being not both 0.

Now we know that the meromorphic functior

$$R(\wp, q) = g(z)$$

cannot vanish identically in z. Determine the number s such that all derivatives $g(z), g'(z), \ldots, g^{(s-1)}(z)$ vanish at the m points $z=z_1, \ldots, z_m$, but

$$\gamma = g^{(s)}(z_k) \neq 0,$$

for some $k=1,\ldots,m$. Plainly, $s \geq n$. The number γ is an integer in K, so that

$$|\mathfrak{N}(\gamma)| \geq 1. \tag{132}$$

On the other hand, because of (130),

$$\overline{|\gamma|} = s!\, c_3^{2n+s} r^2\, O(c_5^n n^n) = O(c_6^s s^{2s}). \tag{133}$$

It remains to find an appropriate upper estimate of $|\gamma|$ itself. Let ν be a positive rational integer which shall tend to ∞ with n, and define

$$A(z) = \prod_{g_1, g_2 = -\nu}^{\nu} (z - g_1\omega_1 - g_2\omega_2)^{3r},$$

$$B(z) = \prod_{l=1}^{m} (z - z_l)^s,$$

where ω_1, ω_2 are basic periods of $\wp(z)$. The function

$$f(z) = \frac{A(z)}{B(z)}\, g(z)$$

is <u>regular</u> in the parallelogram F_ν defined by

$$z = x_1\omega_1 + x_2\omega_2$$

$$(-\nu - \tfrac{1}{2} \leq x_1 \leq \nu + \tfrac{1}{2};\ -\nu - \tfrac{1}{2} \leq x_2 \leq \nu + \tfrac{1}{2}).$$

Choose n already so large that $z_1, \ldots, z_m$ are inner points of F_ν. On the frontier of F_ν we have the following estimates, because of (125) and the periodicity of $\wp(z)$:

$$g(z) = r^2(c_7\nu)^r\, O(c_5^n n^n),$$

$$A(z) = O\left((c_8\nu)^{3r(2\nu+1)^2}\right) = O\left(c_9^{r\nu^2}\nu^{27r\nu^2}\right),$$

$$\frac{1}{B(z)} = O\left(\left(\frac{c_{10}}{\nu}\right)^{ms}\right),$$

hence

$$f(z) = O(c_{11}^s c_9^{r\nu^2} n^n \nu^{28r\nu^2 - ms}).$$

Furthermore, we have at the point $z = z_k$ of F_ν

$$f(z_k) = A(z_k) \lim_{z \to z_k} \frac{g(z)}{B(z)} = \frac{A(z_k) g^{(s)}(z_k)}{s! \prod\limits_{\substack{l=1\\ l \neq k}}^{m} (z_k - z_l)^s},$$

$$\frac{1}{A(z_k)} = O(c_{12}^n), \qquad \prod_{\substack{l=1\\ l \neq k}}^{m} (z_k - z_l)^s = O(c_{13}^s).$$

The maximum property of the absolute value of an analytic function now yields the estimate

$$\gamma = s!\, c_{12}^n c_{13}^s O(c_{11}^s c_9^{r\nu^2} n^n \nu^{28r\nu^2 - ms})$$

$$= O(c_{14}^s c_9^{r\nu^2} s^{2s} \nu^{28r\nu^2 - ms}).$$

Let $r > 112$ and choose

$$\nu = \left[\sqrt{\frac{ms}{56r}} \right] \geq \left[\sqrt{\frac{r}{112}} \right] \to \infty \qquad (n \to \infty);$$

then

$$28r\nu^2 \leq \frac{ms}{2}$$

and

$$\gamma = O(c_{15}^{s}\, s^{2s}\, \nu^{-\frac{ms}{2}})$$

$$= O\left(c_{16}^{s}\, s^{2s}\, \left(\frac{s}{\sqrt{n}}\right)^{-\frac{ms}{4}} \right) = O\left(c_{16}^{s}\, s^{2s-\frac{ms}{8}} \right).$$

Therefore, in view of (128) and (133),

$$\mathfrak{N}(\gamma) = O\left(c_{17}^{s}\, s^{2hs-\frac{ms}{8}} \right) = O\left(c_{17}^{s}\, s^{-\frac{s}{8}} \right) \to 0$$
$$(n \to \infty),$$

in contradiction to (132). It follows that our assumption at the beginning of §3 was false, and $q(z_0)$ is transcendental whenever λ, μ, g_2, g_3, $\wp(z_0)$ are algebraic, λ, μ not both 0.

§5. Some other results

Let us consider some examples for Schneider's theorem. If (ξ_1, η_1) and (ξ, η) are two real points on the ellipse

$$\frac{\xi^2}{a^2} + \frac{\eta^2}{b^2} = 1 \qquad (0 < b < a),$$

then the arc length

$$s(\xi, \xi_1) = \int_{\xi_1}^{\xi} \sqrt{\frac{a^2 - \epsilon^2 \xi^2}{a^2 - \xi^2}}\, d\xi$$

$$\left(\epsilon^2 = 1 - \frac{b^2}{a^2}\right)$$

is an elliptic integral of the second kind, and not a rational function of ξ and η. Therefore s is transcendental if a, b, ξ_1, ξ are algebraic numbers, except in the trivial case that s=0. In particular, the perimeter of the ellipse is transcendental for algebraic a and b. The corresponding results for the circle, a=b, are contained in Lindemann's theorem.

The arc length of the lemniscate

$$(\xi^2 + \eta^2)^2 = 2a^2(\xi^2 - \eta^2) \qquad (a > 0)$$

is given by

$$s(\xi, \xi_1) = a\sqrt{2} \int_{t_1}^{t} \frac{dt}{\sqrt{1-t^4}} \qquad \left(t^2 = \frac{\xi^2 - \eta^2}{\xi^2 + \eta^2}\right).$$

As a function of t, this is an elliptic integral of the first kind. It follows for algebraic a that the arc length on the lemmiscate, between points with algebraic coordinates ξ, η, is transcendental, except in the trivial case s=0. In particular, the perimeter of the lemmiscate is transcendental for algebraic a. Since

$$4 \int_0^1 \frac{dt}{\sqrt{1-t^4}} = \int_0^1 u^{-\frac{3}{4}} (1-u)^{-\frac{1}{2}}\, du$$

$$= B\left(\tfrac{1}{4}, \tfrac{1}{2}\right) = \frac{\Gamma\left(\frac{1}{4}\right)\Gamma\left(\frac{1}{2}\right)}{\Gamma\left(\frac{3}{4}\right)} = (2\pi)^{-\frac{1}{2}}\, \Gamma^2\left(\tfrac{1}{4}\right),$$

the perimeter, for a=1, equals $\pi^{-\frac{1}{2}}\, \Gamma^2\left(\frac{1}{4}\right)$. Therefore the number $\pi^{-\frac{1}{4}} \cdot \Gamma(\frac{1}{4})$ is transcendental. We do not know, however, whether $\Gamma(\frac{1}{4})$ itself is irrational.

All our transcendency proofs made essential use of the fact that the problem can be reduced to the proof of a property of <u>entire</u> functions. This is the reason why the known methods do not work for elliptic integrals of the third kind and not even for integrals of the third kind in the still simpler case of curves of genus 0. For instance, it is not known whether the number

$$\int_0^1 \frac{dx}{1+x^3} = \frac{1}{3}\left(\log 2 + \frac{\pi}{\sqrt{3}}\right)$$

is irrational.

Concerning elliptic functions Schneider discovered another interesting theorem which we shall mention without proof: Let $\wp(z) = \wp(z, g_2, g_3)$ and $\wp^*(z) = \wp(z, g_2^*, g_3^*)$ be two algebraically independent $\wp$-functions with the invariants g_2, g_3 and g_2^*, g_3^*; then at least one of the 6 numbers g_2, g_3, g_2^*, g_3^*, $\wp(z_0)$, $\wp^*(z_0)$ is transcendental, for any z_0 which is not a period for one of the functions.

It is known that $\wp(z)$ and $\wp^*(z)$ are related by an algebraic equation with constant coefficients, if and only if their period lattices are commensurable; this means that

$$(134) \qquad \omega_1^* = \alpha\,\omega_1 + \beta\,\omega_2, \qquad \omega_2^* = \gamma\,\omega_1 + \delta\,\omega_2$$

with rational α, β, γ, δ, where ω_1, ω_2 and ω_1^*, ω_2^* are basic periods. As an application of the theorem, suppose that both

$$\frac{\omega_2}{\omega_1} = \tau , \qquad \frac{g_2^3}{g_2^3 - 27 g_3^2} = j(\tau)$$

are algebraic numbers. We have

$$g_2 = 60 \sum_{\omega \neq 0} \omega^{-4}, \qquad g_3 = 140 \sum_{\omega \neq 0} \omega^{-6},$$

where ω runs over all periods $\neq 0$ of $\wp(z, g_2, g_3)$. Replacing ω by $\lambda\omega$ with suitably chosen constant $\lambda \neq 0$, we may prescribe $g_3 = 1$ in case $j(\tau) = 0$ and $g_2 = 1$ otherwise, so that now g_2 and g_3 both are algebraic numbers. Define

$$g_2^* = \tau^{-4} g_2, \qquad g_3^* = \tau^{-6} g_3,$$

$$(135) \qquad \omega_1^* = \tau\omega_1, \qquad \omega_2^* = \tau\omega_2,$$

then

$$\wp^*(\tau z) = \tau^{-2} \wp(z)$$

and, in particular

$$\wp^*\left(\frac{\omega_2}{2}\right) = \tau^{-2} \wp\left(\frac{\omega_1}{2}\right).$$

Since now the 6 numbers $g_2, g_3, g_2^*, g_3^*, \wp\left(\frac{\omega_2}{2}\right), \wp^*\left(\frac{\omega_2}{2}\right)$ all are algebraic, the theorem implies (134) and, because of (135),

$$(\tau - \alpha)(\tau - \delta) = \beta\gamma ,$$

so that τ is a <u>quadratic</u> irrationality.

This shows that the elliptic modular function $j(\tau)$ is transcendental for every algebraic number τ in the upper half-plane which is not an imaginary quadratic irrationality. On the other hand, it is known from the theory of complex multiplication that $j(\tau)$ is algebraic for any imaginary quadratic τ.

In 1941 Schneider extended one of his results concerning elliptic integrals to abelian integrals on Riemann surfaces $\mathfrak{R}$ of genus $p > 1$: Let w be an abelian integral of the first or second kind and not an algebraic function, and suppose that $L_1, \ldots, L_p$ are p retrosections on $\mathfrak{R}$ which leave $\mathfrak{R}$ connected; then at least one of the p integrals

$$\eta_k = \int_{L_k} dw \qquad (k=1,\ldots,p)$$

is transcendental; in particular, there exists on $\mathfrak{R}$ a closed curve L such that

$$\eta = \int_L dw \tag{136}$$

is transcendental. It should be mentioned that the last formula on p.126 of Schneider's paper is wrong; however, it is not hard to complete the proof by using Cauchy's formula only for one single variable.

An interesting example is provided by

$$w = \int x^{\alpha-1}(1-x)^{\beta-1}dx$$

where $\alpha, \beta, \alpha+\beta$ are rational, but not integral. Then for <u>any</u> closed curve L the corresponding period η in (136) takes the form $\rho\, B(\alpha,\beta)$, where ρ is a number in the cyclotomic field generated by $e^{2\pi i\alpha}$ and $e^{2\pi i\beta}$.

Therefore Schneider's result implies that the number

$$B(\alpha,\beta) = \frac{\Gamma(\alpha)\Gamma(\beta)}{\Gamma(\alpha+\beta)}$$

is transcendental, whenever $\alpha, \beta, \alpha+\beta$ are rational and non-integral. It follows from the transcendency of π that the case $\alpha+\beta = 1, 2, \ldots$ is no real exception. Choosing $\alpha = \beta$ we see that at least one of the two numbers $\Gamma(\alpha)$, $\Gamma(2\alpha)$ is transcendental for all rational non-integral α. But we do not know, for instance, whether $\Gamma(\frac{1}{3})$ is irrational.

As a geometric application of the result concerning $B(\alpha,\beta)$, consider in the complex z-plane the curve consisting of all points z with the property that the product of the n distances $|z-a\epsilon^k|$ from z to the vertices $a\epsilon^k$ $(k=1,\ldots,n;\ \epsilon = e^{\frac{2\pi i}{n}};\ a > 0)$ of a regular polygon equals a^n. For n=1 and n=2 we obtain circle and lemniscate. A simple calculation shows that the perimeter of the curve has the value

$$s = 2^{\frac{1}{n}}\, a\, B(\tfrac{1}{2}, \tfrac{1}{2n});$$

therefore the ratio s/a is transcendental. Another example is provided by the domain

$$|x^{\frac{1}{\alpha}}| + |y^{\frac{1}{\beta}}| < 1$$

in the (x,y) plane, where α, β are positive rational non-integral numbers; its area has the transcendental value

$$J = 4\,\frac{\alpha+\beta}{\alpha\beta}\, B(\alpha,\beta).$$

BIBLIOGRAPHY

Gelfond, A., Sur les nombres transcendants, C. R. Acad. Sci. Paris, 189, p. 1224-1226 (1929); Sur le septième problème de Hilbert, Bull. Acad. Sci. U.R.S.S. (7), p.623-640 (1934).

Hermite, Ch., Sur la fonction exponentielle, Oeuvres III, p.150-181.

Hilbert, D., Ueber die Transcendenz der Zahlen e und π, Math. Ann. 43, p.216-219 (1893).

Hille, E., Gelfond's solution of Hilbert's seventh problem, Amer. Math. Monthly 49, p.654-661(1942).

Hurwitz, A., Beweis der Transcendenz der Zahl e, Math. Ann. 43, p.220-222 (1893).

Koksma, J. F., Diophantische Approximationen, Ergebnisse der Mathematik und ihrer Grenzgebiete 4 (1936).

Kus'min, R. O., Ob odnom novom klasse transcendentnyh čisel, Bull. Acad. Sci. U.R.S.S. (7), p.585-597 (1930).

Lambert, J. H., Mémoire sur quelques propriétés remarquables des quantités transcendantes circulaires et logarithmiques, Opera mathematica II, p.112-159.

Legendre, A. M., Eléments de géometrie, Note IV.

Lindemann, F., Ueber die Zahl π, Math. Ann. 20, p.213-225 (1882).

Loewy, A., Über Matrizen und Differentialkomplexe, Math. Ann. 78, p.1-51 (1918).

Maier, W., Potenzreihen irrationalen Grenzwertes, J. reine angew. Math. 156, p.93-148 (1927).

Schneider, Th., Transzendenzuntersuchungen periodischer Funktionen I., J. reine angew. Math. 172, p.65-69 (1935); Arithmetische Untersuchungen elliptischer

Integrale, Math. Ann. 113, p.1-13 (1937); Zur Theorie der Abelschen Funktionen und Integrale, J. reine angew. Math. 183, p.110-128 (1941).

Siegel, C. L., Über einige Anwendungen diophantischer Approximationen, Abh. Preuss. Akad. der Wissensch., Phys. - math. Kl., Jahrg. 1929, Nr. 1, 70p.

Stridsberg, E., Sur quelques propriétés arithmétiques de certaines fonctions transcendantes, Acta math. 33, p.233-292 (1910).

Weierstrass, K., Zu Lindemann's Abhandlung,"Über die Ludolph'sche Zahl", Mathematische Werke II, p.341-362.

PRINCETON MATHEMATICAL SERIES

Edited by Marston Morse and A. W. Tucker

1. The Classical Groups
   By HERMANN WEYL .... 314 pp. $5.00
2. Topological Groups
   By L. PONTRJAGIN (translated by Emma Lehmer) .... 310 pp. $5.00
3. An Introduction to Differential Geometry
   By LUTHER PFAHLER EISENHART .... 316 pp. $3.50
4. Dimension Theory
   By WITOLD HUREWICZ and HENRY WALLMAN .... 174 pp. $3.00
5. Analytical Foundations of Celestial Mechanics
   By AUREL WINTNER .... 460 pp. $6.00
6. The Laplace Transform
   By DAVID VERNON WIDDER .... 416 pp. $6.00
7. Integration
   By EDWARD J. McSHANE .... 400 pp. $6.00
8. Theory of Lie Groups
   By CLAUDE CHEVALLEY .... 288 pp. $3.00
9. Mathematical Methods of Statistics
   By HARALD CRAMÉR .... 570 pp. $6.00
10. Several Complex Variables
    By S. BOCHNER and W. T. MARTIN .... 216 pp. $4.00
11. Introduction to Topology
    By S. LEFSCHETZ .... 226 pp. $4.00
12. The Topology of Surfaces and their Transformations
    By JAKOB NIELSEN and WERNER FENCHEL .... In preparation
13. Algebraic Curves
    By ROBERT J. WALKER .... 210 pp. $4.00
14. The Topology of Fibre Bundles
    By NORMAN STEENROD .... 232 pp. $5.00
15. Foundations of Algebraic Topology
    By SAMUEL EILENBERG and NORMAN STEENROD 342 pp. $7.50
16. Functionals of Finite Riemann Surfaces
    By MENAHEM SCHIFFER and DONALD C. SPENCER .... In press
17. Introduction to Mathematical Logic, Vol. I
    By ALONZO CHURCH .... In press

PRINCETON UNIVERSITY PRESS
PRINCETON, NEW JERSEY

Milton Keynes UK
Ingram Content Group UK Ltd.
UKHW030742140924
448309UK00001B/35

9 780691 095752